T0103800

Global Sustainable and Healthy Ecosystems, Climate, and Food Systems

Global Sustainable and Healthy Ecosystems, Climate, and Food Systems

DR. ASHOK ALVA

PARTRIDGE

ISBN:	Hardcover	978-1-4828-6436-6
	Softcover	978-1-4828-6435-9
	eBook	978-1-4828-6437-3

Print information available on the last page.

To order additional copies of this book, contact
Toll Free 800 101 2657 (Singapore)
Toll Free 1 800 81 7340 (Malaysia)
orders.singapore@partridgepublishing.com

www.partridgepublishing.com/singapore

CONTENTS

DISCLAIMER:

This book was written during the author's personal time. No resources or official duty time of author's current or past employers were utilized for writing this book. The opinions expressed in this book are solely author's personal views and does not reflect that of his current or past employers. This book is to honor and aim for betterment of all future generations around the world, including the author's children, Aaron and Michael. We have moral duty to do the right thing and mentor the future generation to safeguard abundant and healthy food systems, natural ecosystems, and mitigate any potential negative effects on the climate and environmental resources. This book is also dedicated to the International Year of Soils – 2015. The declaration of this significant event was in recognition of the need to increase worldwide awareness of the soil degradation, desertification, and deforestation to produce increased amounts of food, feed, and fuel to meet the demands of world population and to cater the changing food habits of the emerging economy countries as the income levels rose in the recent years. It is equally important that we need to energize the world population to be sensitive to maintain sustainable natural ecosystems, and environment despite the challenge of need to produce more food, feed, and fuel to cater the increasing world population.

INTRODUCTION

Public knowledge of complexities to maintain delicate balance between protection of sensitive natural ecosystem and maintaining vibrant agricultural enterprise, from farm to table, is rather limited. Furthermore non-agricultural population has limited appreciation of related food safety and security, and nutrition essentials. This knowledge gap is widening in the recent years, particularly in the western countries where there is an abundance of high quality and safe food at very low cost. To some extent this is the result of ineffective public relations by the agricultural community, including the leadership, farming communities, as well as research and extension personnel, in general. Agricultural and ecosystems leaders must make an effort to look at the big picture issues affecting agriculture, food systems, and natural resources, to promote our profession, reach out to related industries to build strength on strength, to enhance our credibility, and impact. We often tend to talk to those within a close circle, failing to recognize the need to get outside our comfort zone to convey our message to the general public. Contributions of innovations in agriculture and ecosystem impact multiple aspects of food system for all citizens, nutrition and food safety, international food security, protection of natural resources, improving soil quality for future generations, sustainability, and adapting to climate change. Despite such diverse areas of impacts of agriculture and ecosystem issues and contributions which practically touch every citizen's day to day life, it is surprising the low level of recognition and support directed to these sectors unlike what is truly deserved.

The American agriculture is highly efficient as evident from only <2 percent of population directly engaged in agricultural production yet capable to produce abundance of high quality food at low price to feed the rest of the US population and also support a robust export. However, we must prepare for the future as the world population increase rapidly and estimated to attain nine billion within the next few years. Although much of the population increase is in Asia and Africa, with globalization of commodity markets this will have an impact on the food systems in the US as well. Therefore, we have to develop strategies to increase food production within the available natural resources, land and water, making sure that the production practices will not result in rapid degradation of these resources. This book is intended to be as a tool to educate the agricultural, non-agricultural, rural, and urban audience as well as decision makers at all level on major big picture challenges facing the agriculture and ecosystems to cope with various factors that are impacting the food system, healthy natural ecosystems, and environmental sustainability.

MAJOR CHALLENGES

1

Increase in world population

The current projections suggest that the world population will attain nine billion in a few years from the current population of about 7+ billion.

This represents almost 2 billion more population in just a few years. This estimate may vary depending on the source of prediction. However, a big increase in world population is imminent within very near future. The world population was 1 billion in 1804. Using this base population, 118 years were needed to increase the population by another billion, i.e. total 2 billion. The subsequent each 1 billion increases steadily required less number of years, i.e. 37 (2-3 billion), 15 (3-4 billion), 13 (4-5 billion), 12 (5-6 billion), and 11 (6-7 billion) years. This graph clearly shows larger the base population contributes to faster unit increase in population.

For a long time, especially in the developed countries, the general perception has been that we have no need to increase food production. With 2 billion more mouths to feed in just a few years it is a wakeup call to those in the agricultural sector to reconsider were we stand on the need for sustained increase in food production. Majority of population increase occurs in the developing countries. This in fact is due to a very large base population and also due to other societal and/or educational differences.

With increasing globalization the ripple effect of any projected shortage of food in any part of the world will impact the entire world immediately. Furthermore, with increasing disposable income of the emerging economy countries, the population in these countries will be demanding high quality food and often increased proportion of meat based diet as compared to plant based diet. In general, the production inputs per unit of meat are significantly greater than that required for a plant based diet. For example, water and energy required for production of 1 kg of beef is 17 and 73 – fold greater, respectively, than those required for production of 1 kg of corn (Table 1). The change in dietary habits of the large population in emerging economy countries, i.e. leading to more meat based diets unlike the traditional plant based diets, brings about an increasing pressure on the agricultural industry to produce increased quantity of meat based products. This trend, in turn, accelerates the rate of utilization of natural resources.

Table 1. Water and energy required to produce one kilogram (kg) of various foods

Food source	Water (liters)	Energy (k Wh)
Lettuce	130	ND
Potato	250	ND
Apple	700	3.7
Corn	900	0.95
Milk	1100	1.6
Peanuts	3100	ND
Eggs	3300	8.8
Poultry	3900	9.7
Pork	4800	28
Cheese	5000	15
Olive oil	14500	ND
Beef	15500	69

Extracted from: Marrin, D.L. International Journal of Nutrition and Food Science 3:361-369, 2014

2

Resources for agricultural production

Land, water, and nutrients are the major resources needed for agricultural production. The increase in cultivable area has been very modest for the last 30 some years

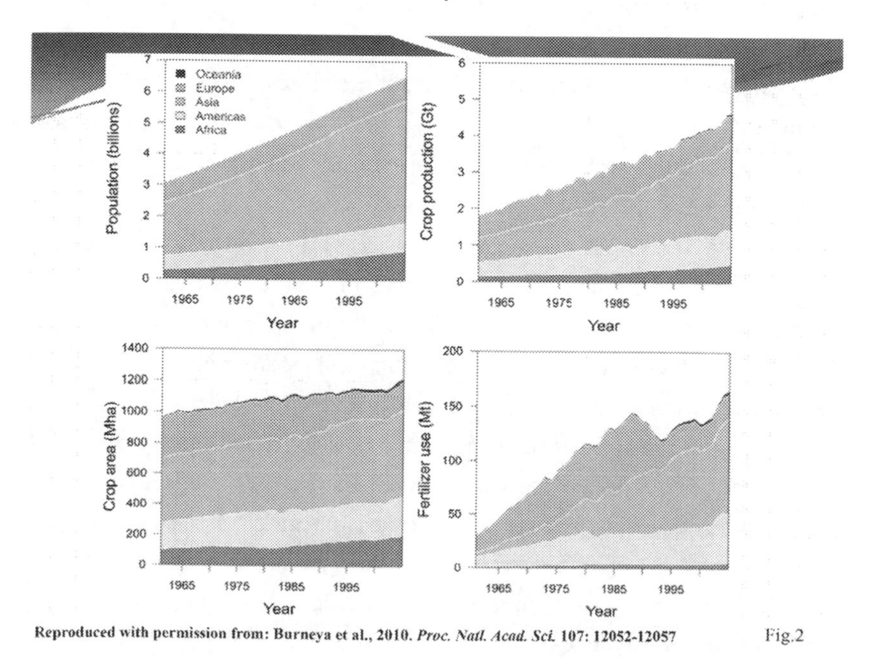

Reproduced with permission from: Burneya et al., 2010. *Proc. Natl. Acad. Sci.* 107: 12052-12057 Fig.2

Trends in population growth, crop production in giga tonnes (Gt), crop area in million hectares (Mha), and fertilizer use in million tonnes (Mt) by different regions of the world.

and the same will be true for the future years as well. In most parts of the world the available cultivable land area has been already utilized to the maximum extent. There is a need to maintain a healthy balance between different land use patterns, including the native preserves, forest land, recreational land use etc. Therefore, increasing the cultivable land area is not a viable option to increase total food production to meet the needs of increasing population.

Soil degradation is already a big problem due to increasing pressure to produce more food. A recent study by the Rothamsted Research station in England shows that if the current intensive production continues, by 2050 the soils productivity will be about 30% less than their current level. If the current population growth continues, by 2050 nearly 70 percent of the earth's surface area will need to be used for agricultural production to meet the food needs of the increasing population as compared to 40 percent being used now. This calls for the urgency for introduction of improved soil management to mitigate the continued soil degradation. The estimate by the United Nations Food and Agriculture Organization (FAO) reveals that 25 and 8 percent of the agricultural lands are highly and moderately degraded, respectively.

Likewise, the water resources are also used to the maximum extent for agriculture purpose. Therefore, despite demonstrated increase in production per unit land area under irrigated production system as compared to that under dryland production system, there is very little prospects of increasing the irrigated land by bringing the dry land under irrigation because of severe limitation of additional water that can be made available for agriculture. As the population continues to increase there are greater competing demands for available fresh water. In some cases, the current agricultural allocation of water may be diverted to other uses, thus, brings additional restrains to increase agricultural production.

Fertilizers are the main sources of plant available nutrients. The organic amendments such as animal manures, sewage sludge, and other plant residues can be used as nutrient sources but the nutrients from these sources are slowly available. Fertilizer use has increased rapidly since the mid 1900's till current. The greatest rate of increase is in the Asian continent due primarily to the pressing needs for increased food production to meet the needs of increasing population. Government subsidy in some Asian countries facilitate relatively low fertilizer price which results in over use of fertilizer in some cases in the hope of increased production. Although this has contributed to increased food production at nominal increase in input costs, if this trend continues there will be some negative impacts. Fertilizer production is very expensive and to some extent resources for fertilizer production are somewhat limiting. For example, phosphorus fertilizers are produced using rock phosphate. At the current rate of usage of phosphate fertilizers the estimates indicate that the world supply of phosphate rock may be exhausted in 60+ years. This can be a serious limitation for future food production that is required to be increased at an accelerated rate to face the challenge of feeding the rapidly increasing world population. Therefore, there is pressing need to explore alternate sources of phosphorus and/or develop new cultivars with increased phosphorus uptake efficiency to ensure adequate food production when the phosphorus reserve is depleted. There are various plant physiological mechanisms enable tapping increased quantity of phosphorus from soils and/or transformation of unavailable forms into plant available forms. Furthermore, genetic modification of crops for increased nutrient uptake or use efficiency can be explored to develop new crop varieties to adapt to low nutrient availabilities, yet produce high yields.

3

Food vs. Fuel competition

The US grain production capacity has increased only marginally since the early 1980s.

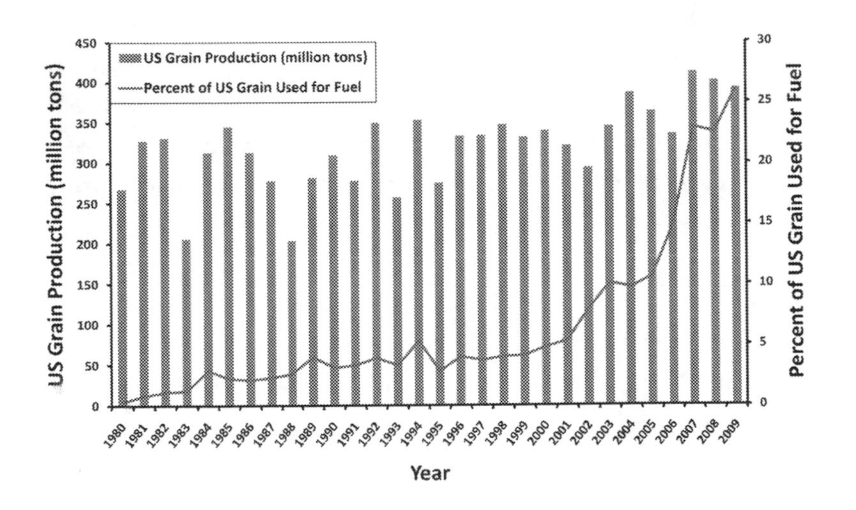

Created using the data from: USDA – National Agri. Statistics Service Fig.3

Trends in US grain production and percent of grain used for biofuel from 1980 to 2009.

Until 1980, the grain production was primarily used for food and animal feed. However, due to increase in demand for fuel, and to enhance energy security, during the recent years there has been a greater emphasis on increasing the domestic biofuel production. This has resulted in increasing proportion of grains being used for biofuels. In 2009, about 26 percent of total US grain production was used for fuel production, which has further increased to 40 percent in most recent estimates. This in turn has impacted the available supply of grains for food and feed. As more and more corn is used for production of ethanol, other grains such as wheat and sorghum have been used as animal feed to overcome the gap in animal feed supply caused by diverting corn for ethanol production. This has contributed to an increase in cost of all animal products including milk, eggs, and meat. In addition the overall increase in cost of fuel in the late 2000s also contributed to an increase in cost of food production and transportation of produce to the market. Although this factor is generally understood by most consumers the impact of food vs. fuel competition for available grain production is not fully appreciated by most consumers.

4

Feeding the World

We often hear about the case being made that US farmers (<2 percent of the total US population) are so efficient in their farming system that they not only feed the US population but also support a strong export market to several countries in the world to feed rapidly increasing world population. This is a rather tall order challenge considering shrinking in US farmers. Therefore, farmers need new tools to produce more per unit area. These tools include improved cultivars which are efficient to produce more with less input, improved chemicals to manage pests and diseases, and precision technologies for various management operations for crop production as well as harvesting, processing, transportation, and efficient marketing of the produce to minimize wastage of the produce, particularly for fruits and vegetables which have very short shelf life.

To some extent these pressures have resulted in clear choice of large scale corporate farming at the expense of small family farming. This has been the clear trend for the past several years in the US, because only the corporate farming can withstand the risks involved in farming and can transition into more efficient systems to adapt improved technologies and practices.

There are various counter arguments for corporate farming. These are considered as environmentally unsustainable and are considered as supporting big business such as major chemical and seed companies. We must also emphasize the fact that emerging economies of the world have gradually adapted improved technology to enhance optimal production efficiencies while maintaining environmental sustainability. Case in point is rapid increase in greenhouse production of vegetables in China, to the tune of 4.7 million hectares. These are very low cost greenhouse construction and management to produce very high yields on a unit land area to meet the increasing demand from growing population for high quality vegetables. Understandingly this resulted in some environmental challenges with respect to increased accumulation of nutrients in small land area, particularly phosphorus, due to the use of high rates of animal manure in addition to inorganic fertilizers. This has implications on degradation of soil quality as well as potential contamination of surface water by excess phosphorus runoff. Research is in progress to overcome the environmental challenges to maintain sustainability, while ensuring continued production of vegetable for increasing population.

5

Disposable income of BRIICS countries

BRIICS is an abbreviation used to include the countries with emerging economy in the world, i.e., Brazil, Russia, India, Indonesia, China, and South Africa. During the recent years, the economy of these countries has increased rapidly, in part due to globalization of trade. This in turn translates into greater disposable income of these people unlike that in the past.

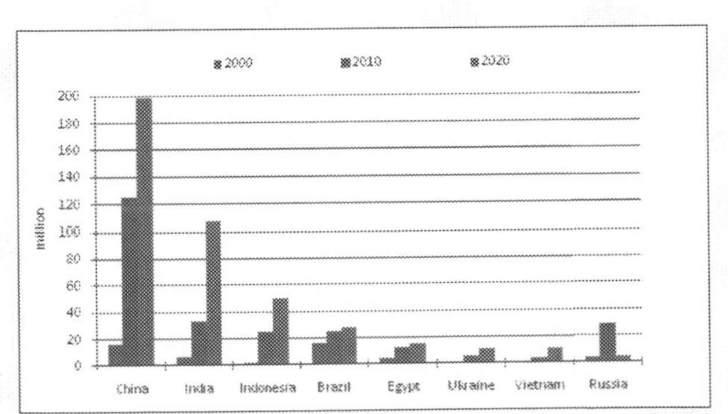

Source: Euromonitor International from national statistics. Note: Data for 2010 and 2020 are forecasts.

Changes in households with annual disposable income range of $5000 to $15000 from year 2000 to 2010, and predicted for year 2020 for several emerging economy countries of the world.

China is leading in this trend. For example, in 2000 there were about 20 million households in China with annual disposable income in the range of $5,000- $15,000. By 2010, the household with the above disposable income group jumped to 120 million and is projected to increase to 200 million by 2020. Similar trend was also evident in other BRIICS countries although not so drastic as that in China. This results in greater buying power of the consumers for various goods including luxury goods. Therefore, the overall production engine gets accelerated around the world and further generates more jobs and accelerates further improvements in economy of the world.

As the buying power of the population in the emerging economy countries improve, they are demanding more nutritious and high quality food and also their food habits gradually change into meat based diet unlike the vegetable based diet. As discussed previously, on a unit weight basis production of meat based food ingredients require more input resources, including water and energy as compared to production of plant based food products (Table 1). This in turn enhances the pressure on resource constraints. As a result, despite the fact that the economic engines in the emerging economy countries are contributing to increased world economic growth, this is exerting increased pressure on already resource strapped agricultural production systems. Awareness of this potential concern is important for all public, since if the current trend in population growth, resources limitation, and competition between food vs. fuel for available agricultural products continue, we can expect continued rise in commodity prices in the future. During the recent 4-5 years agricultural commodity prices have increased at unprecedented rates.

In the developed countries, for example United States of America, among various food related activities, household storage and food preparation accounts for the highest power requirement (32 percent), then followed by agricultural production (21 percent), food processing (16 percent), product transport (14 percent), food service (10 percent), and finally packaging and materials (7 percent) (Table 2). Therefore, nearly 80 percent of the total power requirement for food production through food to the table (i.e. from farm to table), is accountable to all activities except production of food.

Table 2. Power requirements for various food related activities in the United States of America

Activity	BTU/yr (x 10^{12})	k Wh/yr/person (x 10^{3})
Agricultural production	2200	2.4
Product transport	1390	1.5
Food Processing	1680	1.9
Packaging & materials	680	0.75
Food service (retail)	1050	1.2
Household storage & preparation	3250	3.6

Extracted from: Marrin, D.L. International Journal of Nutrition and Food Science 3:361-369, 2014

As expected, the annual food expenditure increased along with annual income growth as evident from the example of the data from China during 2000-2003.

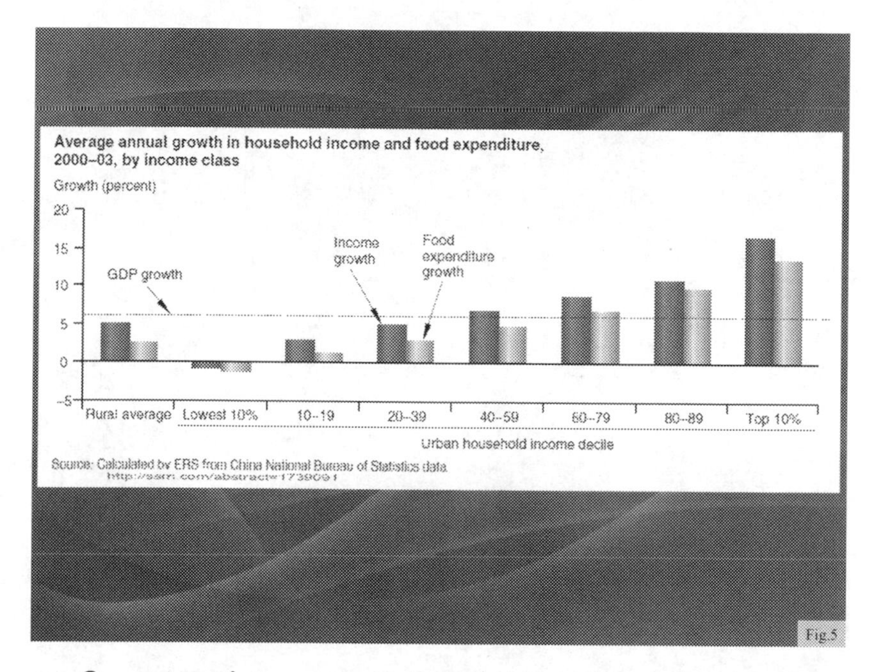

Comparison of average annual household income growth and food expenditure growth from 2000 to 2003 by income class in China

Furthermore this is directly correlated with the household income. The increases in income growth as well as food expenditure growth were greater at the top household income group, as compared to that in the bottom income group. Unfortunately, in the lowest 10 percent of the household income group both income and food expenditure growth declined during 2000-2003. In all other income groups household income as well as food expenditure increased.

The percent of household disposable income spent on food is generally lower in the developed countries as compared to that in the developing countries.

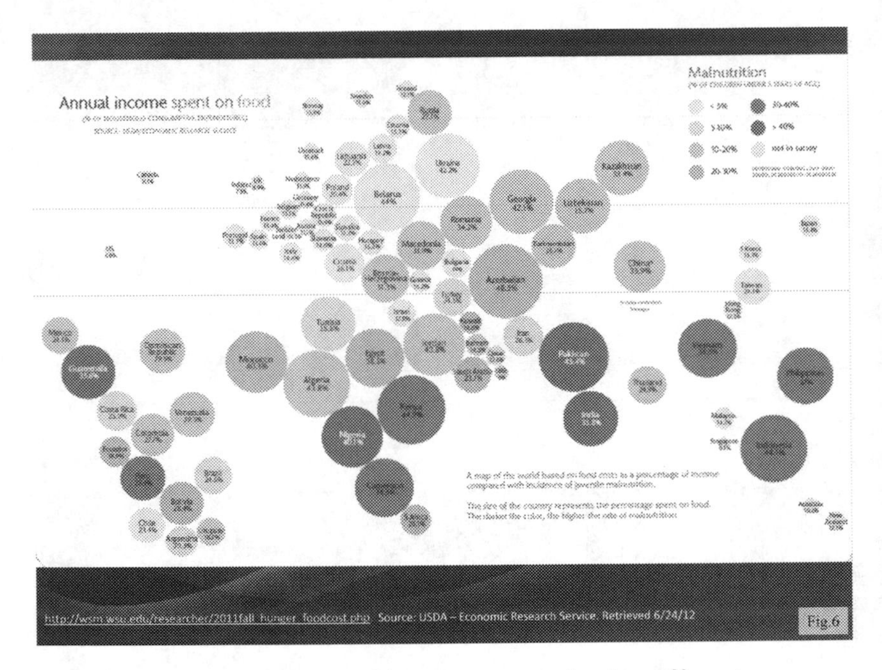

Percentage of annual income spent on food in different countries. Color coding of the circles indicate percent of malnourished children under 5 years.

For United States of America, Canada, most of Western Europe, Japan, South Korea, Australia, and New Zealand the food expenditure as percent of household disposable income is <15 percent. The above percentage is greater, i.e. 20 to 30 percent range, for most of the central and South American countries and some East European countries. Rest of the countries, i.e. in Africa, Asia, and some east European countries food expenditure accounts for 30 to 45 percent of annual household disposable income. This figure also shows high percent mal nourishment of children is directly proportional to the percent food expenses of the total household income. For example, > 40 percent of children below 5 years old are mal nourished in India, Pakistan, Nigeria and Guatemala where percent household income spent on food range from 35 to 45 percent.

6

Irrigated Agriculture

As expected, production per unit area is greater in irrigated agriculture as compared to that in non-irrigated agriculture. In most cases, despite an increase in cost of irrigation infrastructure and cost of pumping water (energy cost), the net return is certainly greater in irrigated farming as compared to that in non-irrigated farming. However, the major limitation for most of the current rain fed cultivated land is lack of available water. The current available water is under a lot of pressure from competing demands. Irrigated agriculture is a major player in the Western US and in Florida.

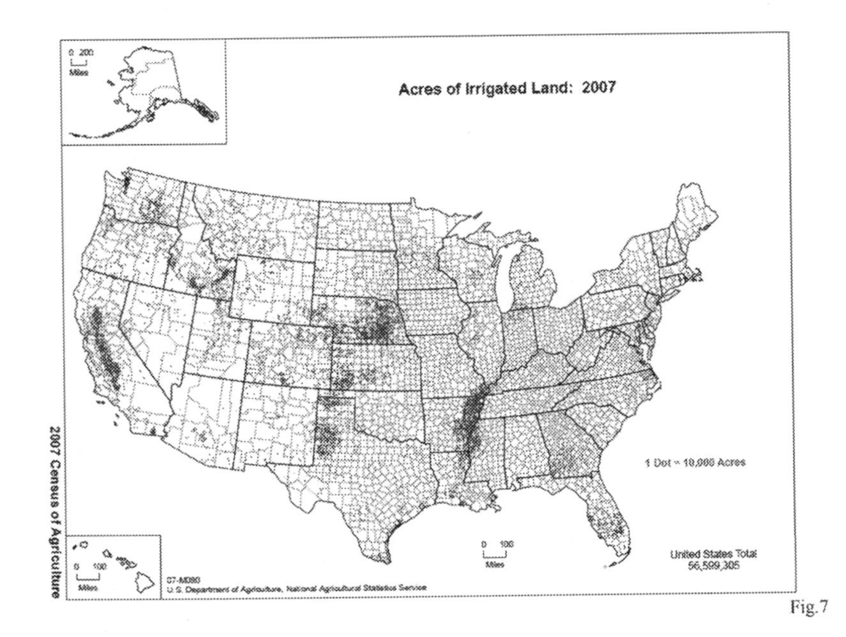

Fig.7

Distribution of irrigated acreage in the United States of America
Each dot represents 10000 acres (1 acre = 0.405 hectare)

Greater than 60 percent of all harvested cropland is under irrigation in most western states and in Florida.

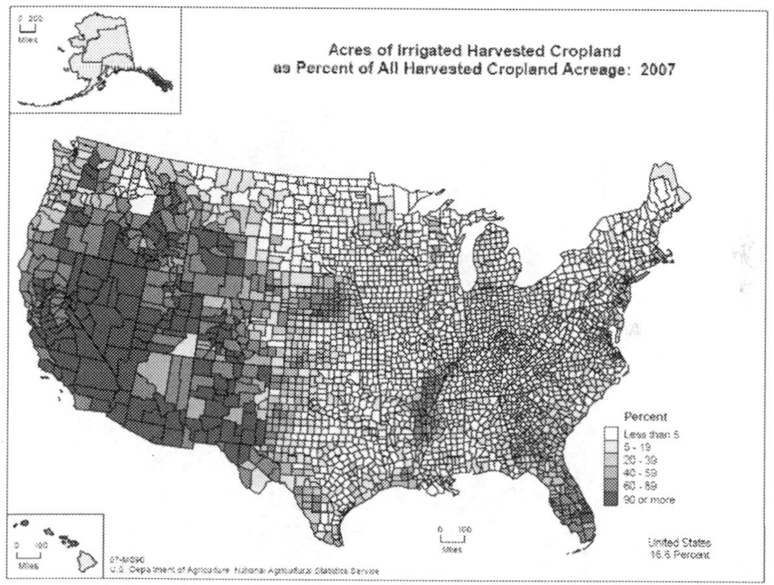

Fig.8

Irrigated cropland as percentage of total acres under cropland in the United States of America. Notice high percent of cropland under irrigation in the western part of the US. In the east, Florida is the only state with high percent of cropland under irrigation.

An evaluation of change in irrigated acreage from 2002 to 2007 reveals that Nebraska showed largest increase in irrigated land with almost 1 million acres added as irrigated land during this period.

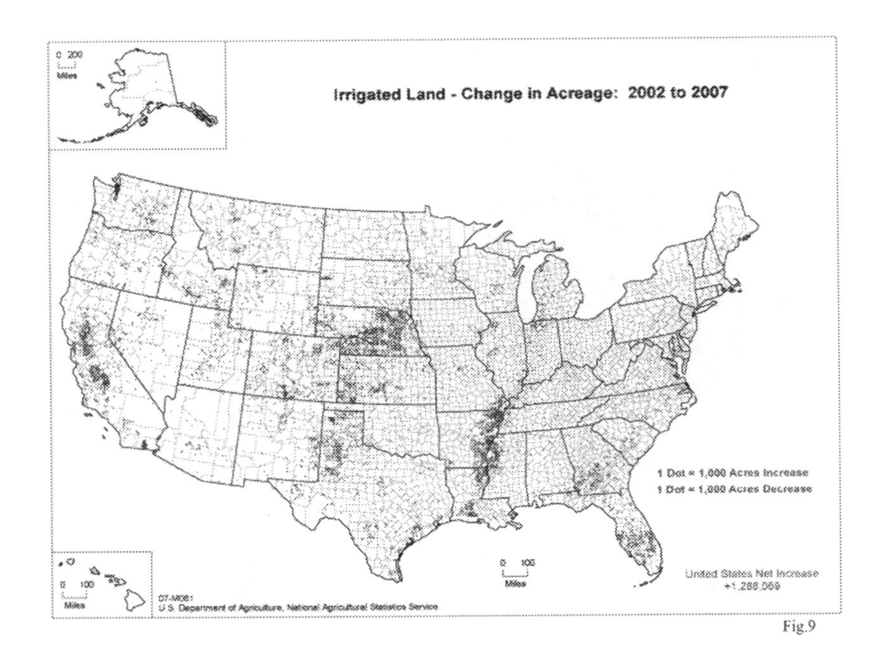

Fig.9

Change in irrigated acreage in the United States of America from year 2002 to 2007. Each dot represents 1000 acres of irrigated land increase (in blue) or decrease (in red).

On the contrary, California, Florida and part of Texas showed significant decline in irrigated land from 2002 to 2007. This could be due to increased competition for available water by urban growth. Despite the proven advantages of high productivity in irrigated agriculture, increasing agricultural land area under irrigation is limited by the following factors: (i) increasing demand for available fresh water to support rapid increase in population and economic growth; (ii) growth in energy sector that is depending on water; (iii) projected impact of climate change on water availability; (iv) potential negative impacts of irrigated agriculture on water quality especially under poor irrigation management; (v) increasing emphasis on water conservation limits issue of new water rights to irrigate drylands to facilitate increased agricultural production; (vi) high cost of irrigation infrastructure for moving water from an area of excess to deficit areas; and (vii) lack of economic incentives for the farmers to invest in high cost of irrigation infrastructures for delivery systems to boost water use efficiency.

However, there is a lot of age old built in loop holes that limits progress in increasing water use efficiency. Case in point is the 'use or loose' provision of water rights in some states. This rule implies that most water rights are already allocated and that it seems virtually impossible to obtain any new water rights. However, the old water rights allocated mandates using the full amount of allocated amount of water per unit land area (stated in the permit) for different crops irrespective of the changes in water requirement of the crop due to climate variations. If the farmers don't use this pre-allocated amount of water, they then loose the right for the amount they don't use. Because of this rule, farmers have no incentive to conserve water, and they apply water regardless of the need of the crop because once they lose water rights they cannot claim it back, in the event that the climate pattern changes and becomes drier requiring more water needed for optimal production.

The practice of application of water in excess of what the crops need, to satisfy the current use or loose provision, contributes to waste of water but also leaching of nutrients and chemicals in the soil profile into below the root zone. The water carrying nutrients and/or chemicals leached below the root zone is not available for root uptake, therefore, will continue to transport down the soil profile and can become the source of groundwater contamination. In addition to wasting precious water resource, irrigation in excess of crop requirement may contribute to negative environmental impacts with respect to deterioration of water quality.

Furthermore, in some states water is free and only cost of irrigation is for energy cost for pumping. The pumping cost is low in cases of low electricity cost, as in the northwest US due to abundance of hydro power. Hydro power is more economical unlike the non-renewable sources such as fossil fuel generated electricity.

7

Total U.S. Cropland

The total US cropland since the early 1900s has remained fairly steady around 330 to 380 million acres.

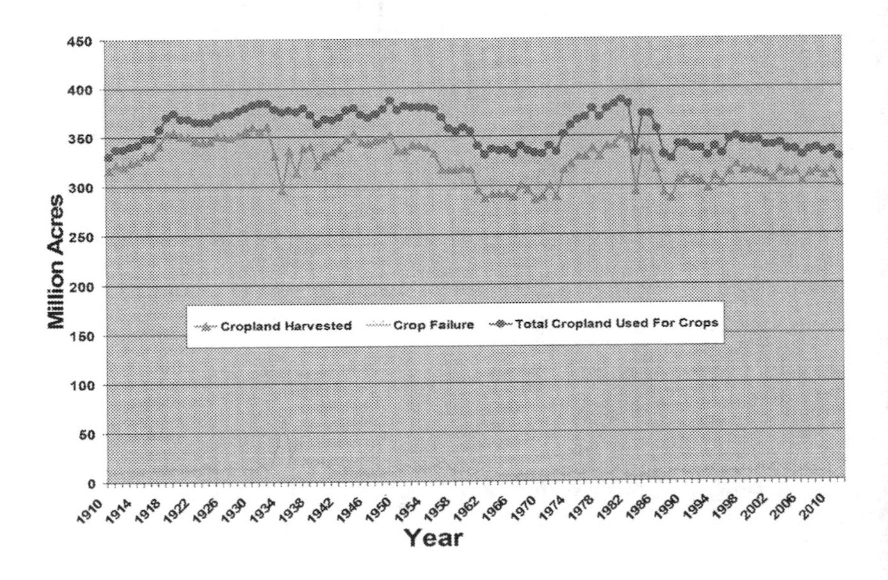

http://www.ers.usda.gov/data-products/major-land-uses.aspx#25962 Fig 10

Total acreage used for cropland, acres of crop failure and the acres of crops harvested in the United States of America from 1910 to current. The trend is remarkably steady over the last 100 years.

Invariably there is a small portion of the crops planted ends up failure, which leaves harvested cropland in the range of 280 to

360 million acres. The area under cropland increased slowly in the 1900's through the 1920 and remained high (360 to 380 million acres) through 1955. This was followed by a decline in acreage to almost 340 million acres until about 1975, and then increased to 380 million acres till 1983. There was a big decline in 1984 which was rebound again to 370 million acres. During the recent decades the cropland acreage remained relatively steady at 330 to 350 million acres. A large portion of this cropland is under corn and soybean each accounts for about 70 to 80 million acres.

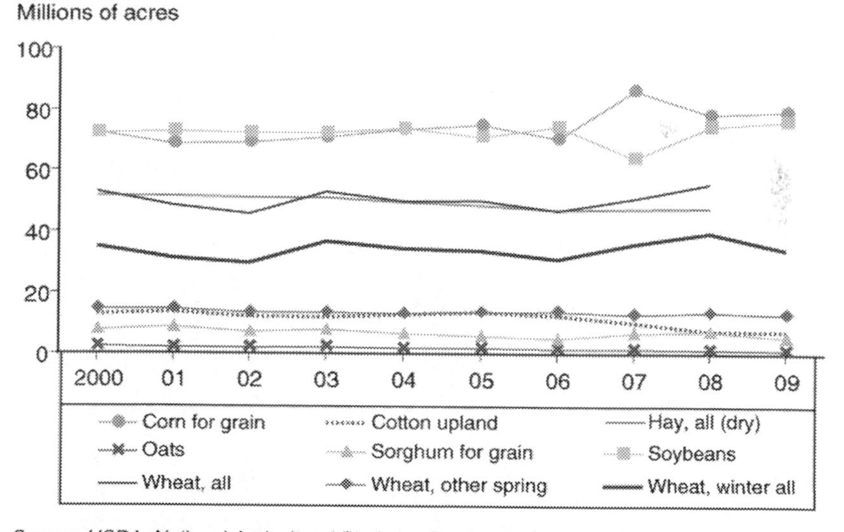

Harvested acreage (in million acres) of major crops in the United States of America since year 2000 to 2009.

This is mainly because corn – soybean is a common rotation in most of Midwest states. This kind of cereal – legume rotation is preferred, unlike the monoculture, to improve soil physical and biological properties and maintain improved soil fertility and productivity. Both of these crops are important food and animal feed sources, therefore, require abundant production to meet the demand.

Recently however, a significant portion of the corn is diverted to production of ethanol. This shortfall of corn for food and animal feed is made up by wheat and grain sorghum. Next in the order of acreage ranking is Wheat and Hay, each occupied an area of about 50 to 55 million acres. Of the total wheat acreage, approximately 35 to 40 million acres is winter wheat, and the rest is spring wheat. This is followed by cotton (10 to 15 million acres), sorghum (10 million acres) and Oats (3 to 5 million acres). Grain crops occupy the largest acreage of the total available arable land.

8

Crop Area vs. Crop Value

Total grain crop production in the US is well over 145 million acres i.e. about 47 percent of total cultivable land area.

GRAIN AND SPECIALITY CROPS

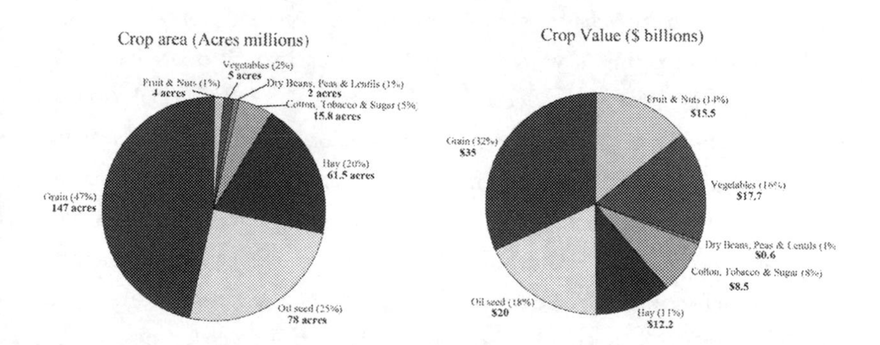

Fig.12

Total crop area (in million acres) and crop value (in billion dollars) across the United States of America for grain crops, hay, oils, and different specialty crops

This is predominantly corn, and wheat and to smaller extent rye and oats. These crops represent major food and feed crops, thus, the need to produce in large quantity. Recently almost > 30 percent of corn produced is utilized for ethanol production. Therefore continuing need to produce the grain crops in large quantities is undeniable. The next in ranking of acreage is oil seed with close to 80 million acres, i.e. 25 percent of available arable land. This is primarily soybean which is grown as a rotation crop with corn. Grain and oil seed crops combined account for about 72 percent of the total arable land. However, the crop value of these two commodities account for only 50 percent of total crop value in the US. Hay crops acreage is about 62 million acres, i.e. 20 percent of the total cultivable land. However, the economic value of hay crop is only 11 percent of total value from all crops. Cotton, sugar, tobacco crops account for 5 percent of the total arable land with 8 percent of total crop value.

The most striking statistics is that area under fruits, nuts, vegetables, including edible legumes combined referred to as 'specialty crops', accounts for only 4 percent of the total arable land, but these specialty crops contributes to 31 percent of the total crop value, this is very similar to the crop value of the grain crops which by the way occupies 47 percent of the total arable land. This signifies the greater economic value to the specialty crops despite using only 4 percent of the total crop land. This has some implication on the Federal support both in terms of subsidy as well as research funding. Farm subsidy is mainly for grain crops and cotton, which combined account for 52 percent of the arable land. The farm subsidy program covers corn, cotton, soybean, wheat, tobacco, dairy, rice, and peanuts. Almost 90 percent of the total subsidy however, goes to the first five commodities. Likewise, most of the Federal research funding goes to the grain crops which occupy very large acreage. To some extent this emphasis is to ensure continued production of adequate quantities of grain crops and soybean, which constitute the major food, and animal feed crops. This philosophy has been seriously contested during the recent years by the specialty crops producers by emphasizing their significant contribution (31 percent) to the total crop value of all commodities, despite requiring only a small fraction of the total arable land (4 percent). This in turn resulted in a new source of research funding specifically earmarked for specialty crops called 'specialty crop research initiative' (SCRI) as a part of the Farm bill.

There has been appreciable shift in distribution of economically significant farming areas from 1950 to 2000.

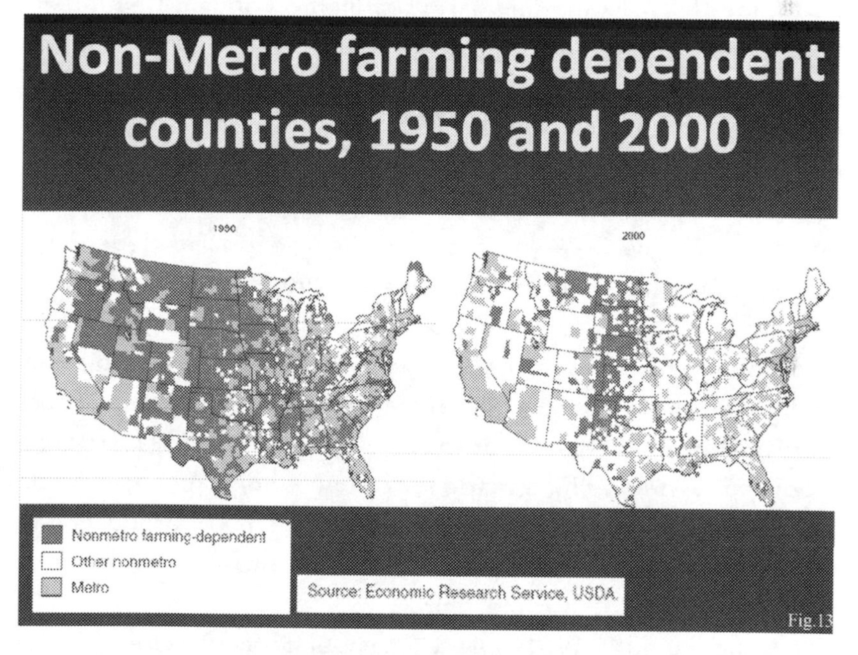

Changes in non-metro farming dependent counties in the the United States of America from 1950 to 2000.

According to USDA- Economic Research Service (ERS), in 1950 Non-metro Farming Dependent County was defined on the basis of at least 20 percent of county income derived from agriculture. In the 1950, the Non-metro Farming Dependent counties were distributed throughout the country, although the major concentration was in the mid-western and western states, except California and Arizona. By 2000, the Non-metro Farming Dependent counties decreased considerably, and it was predominantly in 7 states (Montana, North Dakota, South Dakota, Nebraska, Kansas, Oklahoma, and

Texas). Industrialization and other non-agricultural enterprises as major economic engines contributed to this drastic change. Furthermore, increased efficiencies of agricultural production and rapid expansion of urbanization, contributed to survival of more corporate farms at the expense of small family farms predominantly concentrated in less populated rural areas. During 1998 through 2000, the USDA-ERS definition of non-metro Farming Dependent County was based on ≥ 15 percent of average annual labor and properties earnings derived from agriculture or ≥15 percent of employed residents worked in agriculture related jobs. With increased mechanization, the need for labor force in agriculture declined considerably in the last 2-3 decades. This in part explains the reason for the fewer counties classified as farming dependent by year 2000. It is important to recognize that agriculture continues to play a major role in the US to provide the most nutritious, safe and abundant food products at low price. This is despite an increase in population with greater demand for nutritious food supply. Modern agriculture is mechanized and is highly efficient production system with dependency on low labor force. As a result the population base directly involved in agricultural production is rather small. Therefore, the political power base of the modern agriculture in the US is rather small.

During the early 1900's, the US number of farms were ≥ 6000 which decreased steadily down to ≤ 2000 by 2002, i.e. 63 percent reduction.

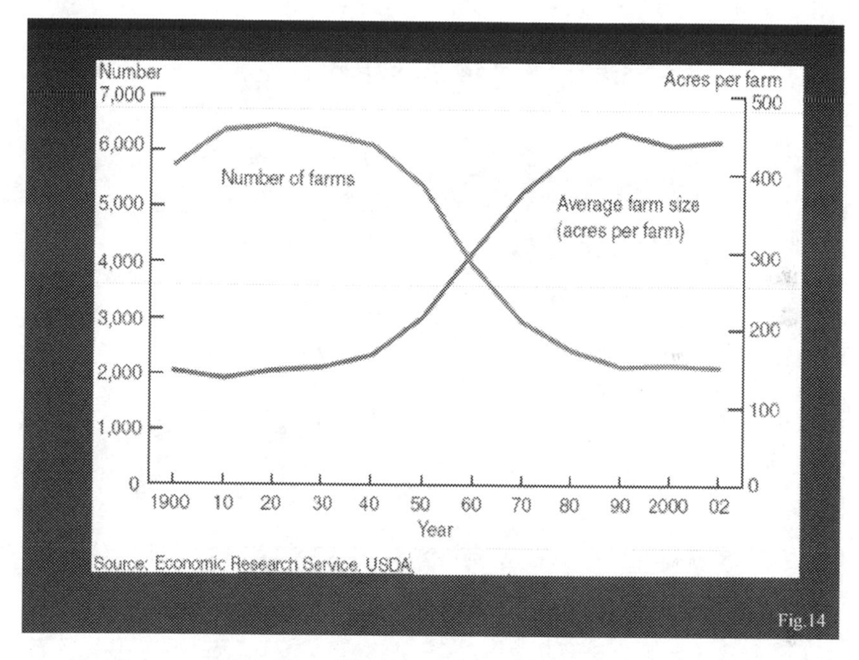

Changes in number of farms in the United States of America and acreage per farm from 1900 to 2002.

During the same period, the average farm size increased from ≥ 150 acres to 450 acres, i.e. 200 percent increase. A close examination of US grain production in the last two decades shows, total grain production increased from about 200 million tons in 1995 to about 370 million tons by 2007 and 2009, (85 percent increase) and subsequently declined to 300 million tons by 2012 (19 percent decrease). During the same period, total soybean production increased from 70 to 100 million tons (43 percent increase), and wheat production remained fairly steady at 70 million tons with some annual fluctuations.

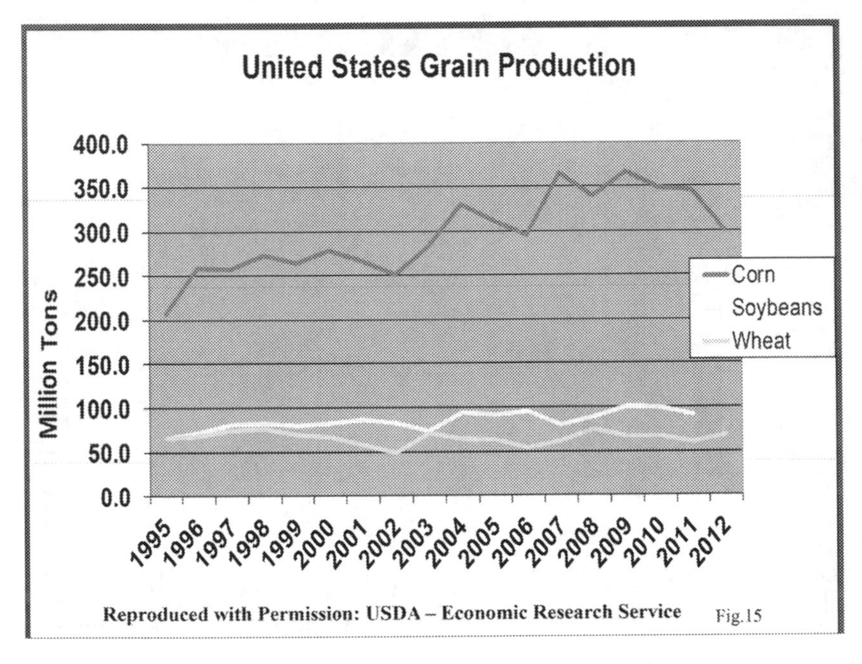

Trend in grain production (in million tons) in the United States of America from 1995 to 2012

Corn is the most widely produced animal feed in the US. It is used as a main energy ingredient in livestock feed. Furthermore, corn is also processed into a multitude of food and industrial product including starch, sweeteners, corn oil, beverages, industrial alcohol, and most recently as ethanol fuel. Approximately 20 percent of corn produced in the US is exported to other countries.

The structure of US farms has gone through drastic changes in the last century. In the early 1900's, US farms were labor intensive, large number of small and diversified farms, with ≥ 5 crops. Approximately 50 percent of the US population was residing in the rural area. During the 21 century the US farms became highly mechanized with minimal labor dependence. The farms became large, specialized (1-2 crops only), and highly productive. As a result the number of farms declined due to competition from the large farms and consolidation of small farms unable to stay economically viable. The rural area population accounted $\leq$ 25 percent of total US population. Most recently, the US population engaged in agriculture declined to $<$ 2 percent of total population. High cost of labor was the driving force for introduction of increased mechanization in farming. However, the small farms are unable to afford high initial investment on machinery. Increasing competition due to globalization of agricultural market along with increased cost of fuel and other inputs for agricultural production resulted in a decline in net returns per unit land. This was the driving force for introduction of increased mechanization in farming. However, the small farms were unable to afford high initial investment on the machinery. This, in turn, forced the small farms go out of business, therefore, resulted in consolidation of small farms into large corporate farms to make it economically viable enterprise. Farming became increasingly risky due to unexpected factors such as weather, fluctuations in global market, and other uncertainties. To sustain such external pressure it was necessary to rely on large scale farming.

9

Recent changes in commodity prices

The commodities prices were relatively stable in the 1990's, but started to increase by 2002, hit a peak in 2008.

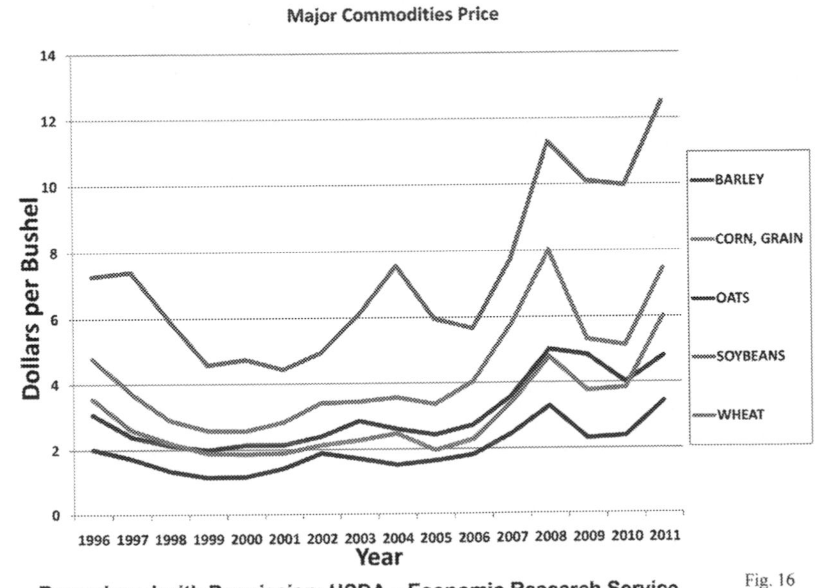

Reproduced with Permission: USDA – Economic Research Service

Fig. 16

Major commodities price (dollars per bushel) in the United States of America from 1996 to 2011. Notice the steep increase in most commodities prices following 2005.

The prices began to slide a little bit and then rebound to another peak by 2011. The 2011 peak price was almost 2-fold greater than that in 1996 for some commodities, i.e. for soybean the price increased from \$7 to \$13 per bushel. Price increase for corn for the similar period was from \$3.5 to \$6 per bushel. During the peak of prices of 2008 and 2012, the overall food grains price increased by almost 300 percent as compared to the base price of 1990-1992.

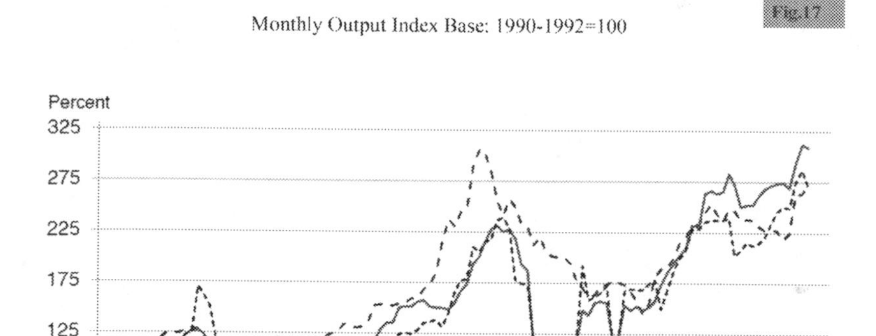

Reproduced with Permission: USDA - Economic Research Service

Percent increase in output index from 2003 through 2013 for food grains, feed grains and hay, and oil seeds. The base index of 100 was for the price in 1990-1992

During the similar period, the prices of oilseeds increased by about 230 percent, and that of feed grain and hay increased almost by 320 percent. The increase in fuel cost and the subsequent increased emphasis on greater targets for biofuels contributed to food and feed vs. fuel competition for grain and hence increased commodity prices.

Similar evaluation of dairy products showed a price increase in 2008 and 2012, of 170 percent as compared to that in 1990-1992.

Fig. 18

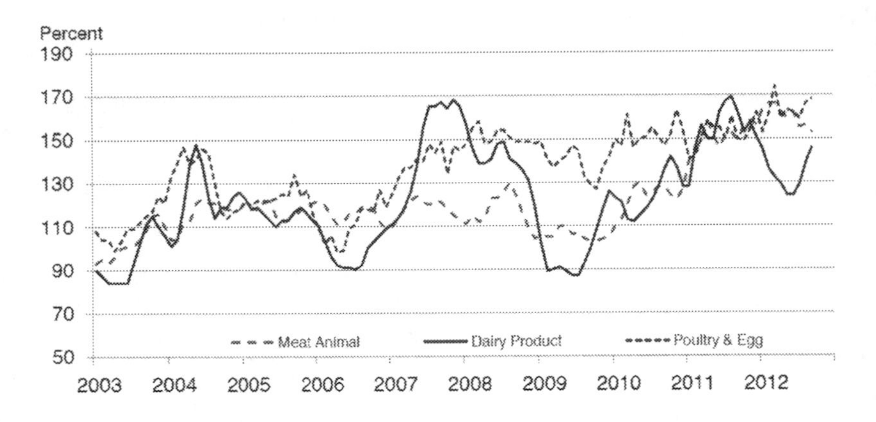

Reproduced with Permission: USDA - Economic Research Service

Percent increase in output index from 2003 through 2013 for meat, dairy products, and poultry and eggs. The base index of 100 was for the price in 1990-1992

Poultry and egg price increased by 150 percent in 2008 and continued to increase up to 170 percent by 2012. The lack of price increase that was seen for most commodities between 2008 and 2012 was not evident for poultry and egg price. This trend was also true for meat price which increased by 130 percent in 2008 and continued up to 170 percent by 2012. The continued price increase for poultry, eggs and meat products through 2008 and 2012 can be, in part, due to increase in feed price due to increasing pressure on grain market due to use of corn for ethanol.

Corn export decreased in 2010 as compared to that in the previous two years, and showed further big decrease in 2011.

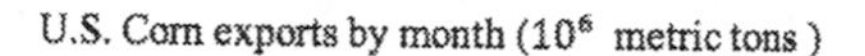

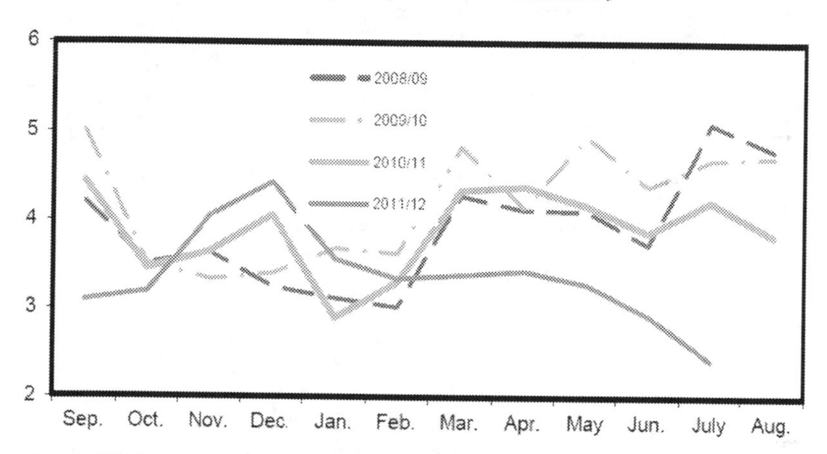

Source: USDC, U.S. Census Bureau, http://www.usatradeonline.gov/.

Fig. 19

Corn export (in million metric tonnes) from the United States of America by each month for 2008 through 2011

This again explains an increasing proportion of corn is diverted to fuel production. The 2008 price increase was much greater for corn, wheat and soybean as compared to that for cotton.

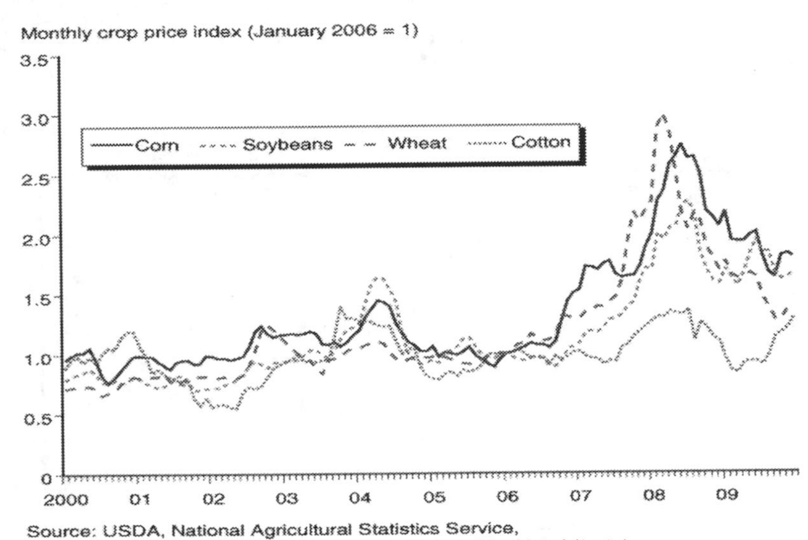

Fig. 20

Monthly crop price index of different agricultural commodities from 2000 to 2010, based on index of 1 for 2006 price. Notice the steep increase in price index post 2006 for corn, soybean, and wheat.

This further substantiates pressure on corn for ethanol production and increased use of wheat and soybean for animal feed to overcome the shortfall.

A number of factors impacted this increase in commodities prices including, increase in energy cost, uncertainties in the future market, population increase which created an increased demand for food, diverting the agricultural products from food/feed into fuel in an effort attain the ambitious targets set for energy security, and overall increased cost of all inputs needed for agricultural production. The prices will hopefully level off, because the rate of commodity price increases seen in the recent years is certainly unsustainable.

10

Increase in cost of agricultural inputs

The commodities price increase is indeed a reflection of rapid increase in cost of agricultural inputs. As explained above the cost of fuel increased rapidly in the early to mid-2000. This has impacted the price of agricultural inputs including fertilizers and chemicals. The prices of anhydrous ammonia, super phosphate, potassium chloride, and diammonium phosphate were 7 to 8 - fold greater in 2008 as compared to the respective prices in 1960.

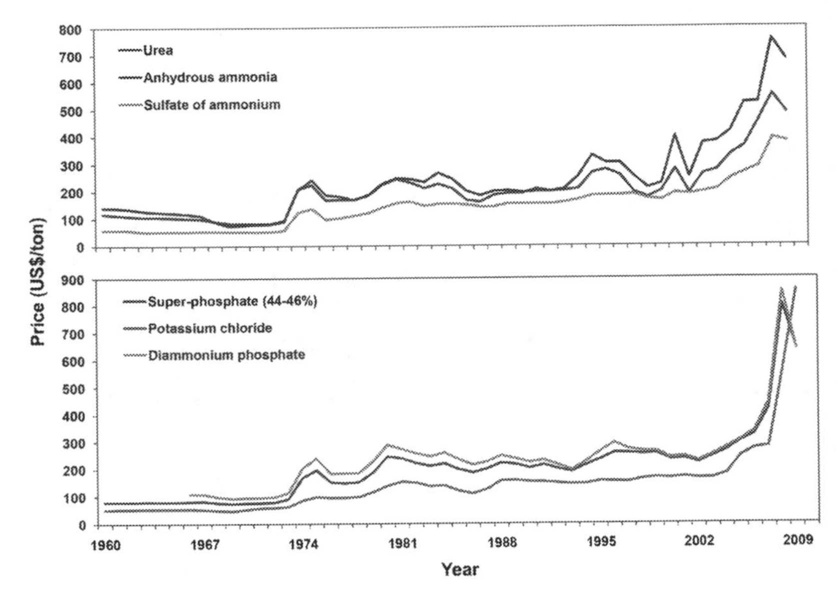

Market price of different fertilizers (US $/ton) from 1960 to 2010. Notice the steep increase in all fertilizers post 2005.

From 1960 to 2002 the increase in price of the fertilizers was 2 to 3- fold. Subsequently, the price increase was rather steep and exceptional. The increases in price of urea, and ammonium sulfate from 1960 to 2002 were 3 to 4- fold. Transportation cost is a major component of fertilizers, since all bulk produced fertilizers must be transported from the production facilities, to the distribution facilities around the country and, in turn, transported to the farms.

11

Emission of greenhouse gases (GHG) from agriculture

In 2005, agricultural practices were responsible for 1.4 to 1.7 giga tons (GT) of carbon (C) emissions which was 10 to 14 percent of the total anthropogenic greenhouse gas (GHG) emissions. The main components of agricultural emissions of GHG are: (i) Nitrous oxide (N_2O) released from soils associated with N fertilizer application (38 percent); (ii) Methane (CH_4) from livestock enteric fermentation, and CH_4 and N_2O from manure management (38 percent); (iii) CH_4 from cultivation of lowland rice (11 percent), and (iv) from burning agricultural residues (13 percent).

In Nature, CO_2, CH_4, and N_2O are continuously emitted to and removed from the atmosphere. Anthropogenic activities, however, can cause additional quantities of these greenhouse gases to be emitted or sequestered, thereby changing their global average atmospheric concentrations. Particularly, poor water and fertilizer management accelerates the trace gas emissions. For example, over irrigation or flooding can increase N_2O and CH_4 emissions from crop production due to reduced soil conditions which contribute to lower redox potential. In soils CO_2, N_2O and CH_4 are produced or consumed by microbial activities, which are regulated by the soil redox potential (abbreviated as Eh) and the availability of the energy source and the electron acceptors. Natural activities such as respiration by plants or animals and seasonal cycles of plant growth and decay are examples of processes that cycle carbon (C) and nitrogen (N) between the atmosphere and biomass.

Carbon dioxide concentration in the atmosphere increased from approximately 280 parts per million by volume (ppmv) in pre-industrial period to 385 ppmv in 2008,

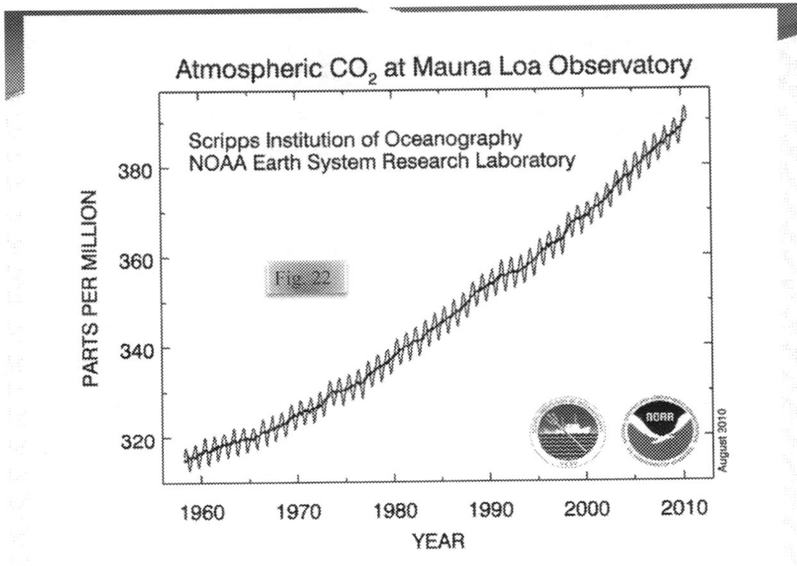

Changes in concentration of atmospheric carbon dioxide from 1960 to 2010.

a 37.5 percent increase. The Inter-governmental Panel on Climate Change (IPCC) concluded that the present atmospheric CO_2 increase is caused by anthropogenic emissions of CO_2. The global temperature has steadily increased from 1880's through 1960, followed by much accelerated increase since 1960's to present.

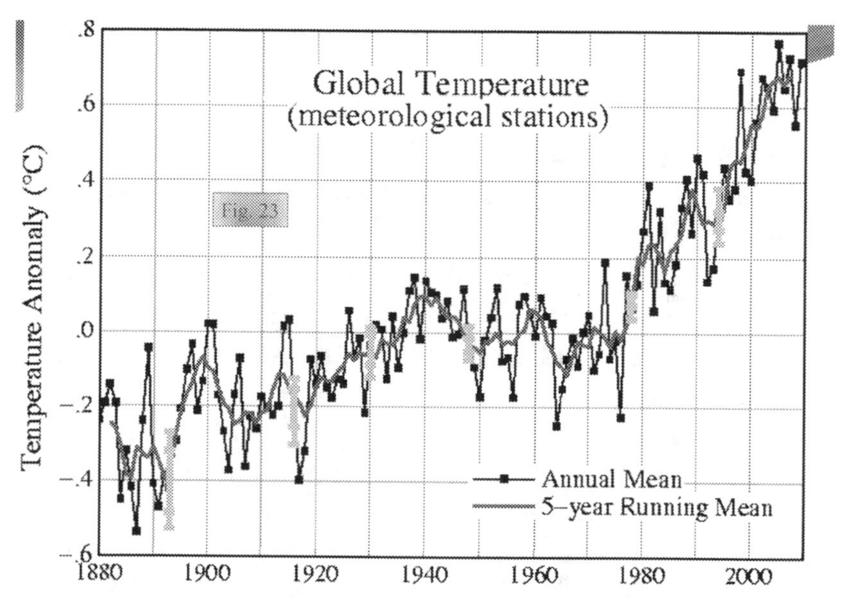

Reproduced with permission from: Hansen et al., 2006, *Proc. Natl. Acad. Sci.* 103: 14288-14293

Changes in global temperature, as annual mean of five years running mean, during 1880 through 2010.

Methane is produced during anaerobic decomposition of organic matter in biological systems. Agricultural processes such as wetland rice cultivation, enteric fermentation in animals, and the decomposition of animal wastes emit CH_4. Atmospheric concentration of CH_4 increased by about 143 percent since 1750, i.e. from its pre-industrial era value of about 722 ppb to 1,741-1,865 ppb in 2007. The Inter-governmental Panel on Climate Change (IPCC) has estimated that about half of the current CH_4 flux to the atmosphere is anthropogenic, from human activities including agriculture and other sources such as fossil fuel use, and waste disposal. The atmospheric concentration of N_2O has increased by 18 percent since 1750, i.e. from its pre-industrial era value of about 270 ppb to 322 ppb in 2007. N_2O is primarily removed from the atmosphere by the photolytic action of sunlight in the stratosphere.

Agricultural activities serve as both sources (trace gas emissions) and sink (carbon sequestration) for greenhouse gases due to the differences in the strategy and technique of crop, fertilizer and water management practices. For instance, emission of CH_4 and N_2O from transplanted flooded rice decreased by using wax-coated calcium carbide as a source of acetylene to inhibit nitrification and CH_4 production. Therefore, modification of agricultural practices can reduce net emissions of GHG. Agricultural activities play a major role as a sink for GHG, particularly CO_2 for biomass production and through changes in tillage or land use including conversion of cropland to grassland.

Biofuels derived from low-input high-diversity (LIHD) crop production systems can reduce C emissions up to 4.4 Mg ha^{-1} yr^{-1} from soils. To achieve maximum positive impact on mitigating climate change, it is desirable to reduce net CO_2 emissions and increase soil C storage referred to as C sequestration. Biochar, a byproduct of bioenergy production by pyrolysis of feedstock, can remarkably increase C sequestration when used as a soil amendment for improving soil quality and increasing soil carbon sequestration. Additionally, reduced tillage practices can significantly reduce GHG emissions through formation of coarse intra-aggregate particulate organic matter.

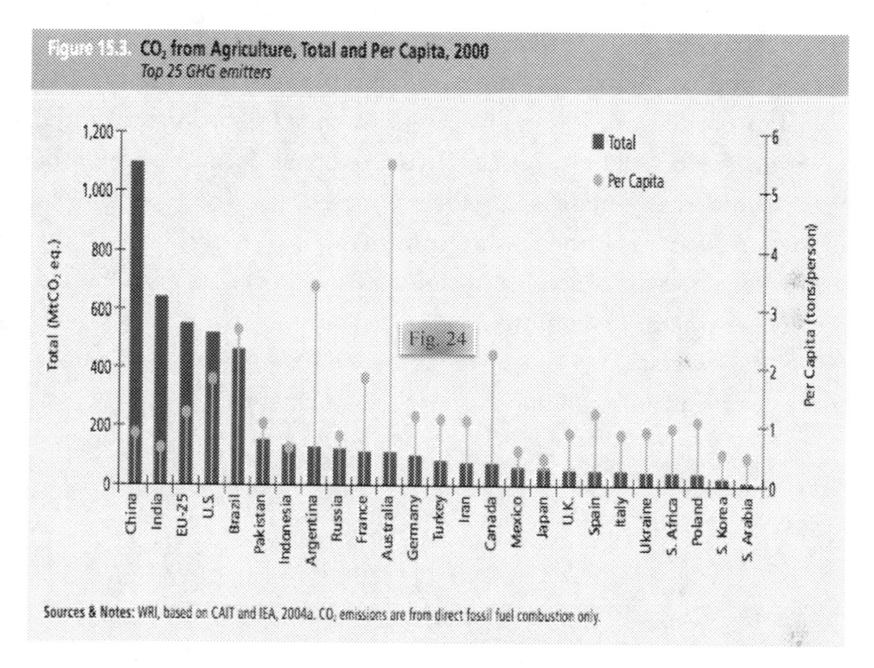

Total and per capita trace gas emissions for different countries.

As shown in the figure above, China and India ranks number 1 and 2 in terms of total quantity of CO_2 emission. However, Australia, Argentina, and Brazil are top ranking countries for high CO_2 emission on per capita basis.

Intensification of agriculture for increased production of high quality and greater nutritious food contributes to greater potential for GHG emissions and associated climate change. Therefore, there is a need to develop sustainable strategies for agricultural production that can mitigate GHG emissions while maintaining high production and quality of crops and animal products to feed the increased population that demands and can afford high nutritious food.

In most cases, environmental sustainability and high production targets are at odd to each other. Likewise, environmental and economic sustainability also don't go together in most cases.

Hence the need to incorporate innovative technologies to develop environmentally sustainable and technically feasible agricultural production practices to increase crop yields, with high produce quality, and nutrition of products with marginal and/or no increase in cost of production. This may appear to be a tall order challenge, however, the one that is required for reduction in degradation of natural resources without a significant increase in food prices. The impact of climate change is wide spread on the entire citizens of the world. Proactive approaches to demonstrate sincere efforts by the agricultural sector to mitigate GHG emission will have a long lasting effect to overcome the negative perception by the general public concerning agricultural impact on the environment.

12

Strategies to mitigate GHG emissions from agriculture

Although the GHG emission from agriculture sector is a small component of the total GHG emissions (< 15%), it is important to take a proactive step in reducing such emissions. Indeed, as it becomes evident from the following discussion, research and demonstrations have thus far shown various options to reduce the agricultural GHG emissions. The case studies discussed below are some of my team research in the US Northwest and other published studies. The following discussion is grouped into five subcategories.

12.1 Reduced Tillage in Irrigated Crop Rotation

The US Pacific Northwest (PNW) includes Washington, Idaho, and Oregon states. The agriculture production regions in these states are predominantly dependent on irrigation since rainfall, is often low. For example, in the Columbia basin region annual precipitation is about 165 mm with only about < 25% of the annual precipitation received during the peak growing season (May through August). The growing season is characterized by dry, hot, long days and cool nights. These climate conditions are conducive for production of high biomass, provided ample irrigation water is available.

Potato is an important crop in the US PNW, which accounts for up to 55 percent of the US total potato production. Generally potato is grown in rotation with wheat and one to two years of corn. Various other rotation crops include alfalfa, onion, sweet corn, mint etc. Potato production requires intensive tillage with up to 8 to 9 heavy equipment passes during the growing season. We conducted a long term study with reduced tillage for potato, with reduction of 3 to 4 equipment passes during the growing season. In addition to tuber yield, tuber quality and soil quality evaluations, GHG emissions were also examined from potato and sweet corn fields under reduced tillage and conventional tillage practices.

Research has shown that reduced tillage decreased CO2 and N2O emissions as compared to those in conventional tillage from corn field. However, the above effects were not evident in potato field. Fertilizer application significantly increased the N2O emission under both potato and corn production as compared to that from the same crop fields without fertilizer application.

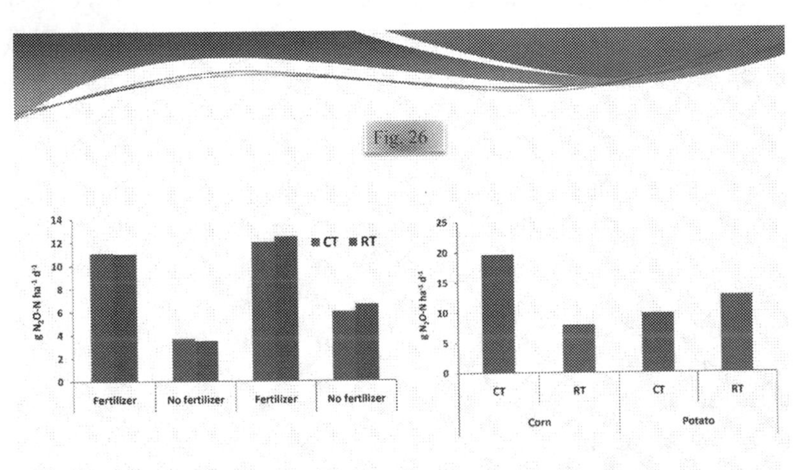

Extracted from: Haile-Mariam et al., 2008. JEQ 37: 759-771

Emission of N2O from potato and corn fields under conventional tillage (CT) and reduced tillage (RT) with or without nitrogen fertilizer application in the US-Pacific Northwest production conditions with Center Pivot irrigation.

12.2 Cover Crops for Carbon Sequestration

Mitigation of GHG can be achieved by two distinct strategies: (i) reduced emission of GHG into atmosphere, and (ii) capture atmosphere CO_2 by way of biomass production and depositing the biomass carbon in the soil, i.e. C sequestration, which provides additional soil quality benefits. The concept of cover crops began with intent to provide vegetation cover to agricultural soils during the off season to reduce soil erosion, conserve soil and water, and enhance nutrient retention. In the US Pacific Northwest, cover crops are sown in September-October that facilitates 20 to 25 days of growth before the onset of killing frost. This vegetation growth provides soil cover over the winter months against heavy wind erosion. During the next growing season this vegetation is plowed in to incorporate the biomass with soil. Thus, the nutrients in the biomass are made available to the next crop following decomposition of the biomass, and mineralization of nutrients into plant available forms. During this process carbon is deposited into the soil that provides improved soil quality and soil health.

Our long term studies have shown under irrigated US-Northwest production conditions, the total biomass from various cover crops (such as Mustard, Sudan grass, wheat and Hairy vetch plus Oats) following sweet corn ranged from 1.4 to 6.1 tonnes per hectare, and 1.3 to 3.5 tonnes per hectare following wheat. The above ranges of biomass contribute 0.6 to 2.4 and 0.5 to 1.4 tonnes C per hectare, respectively.

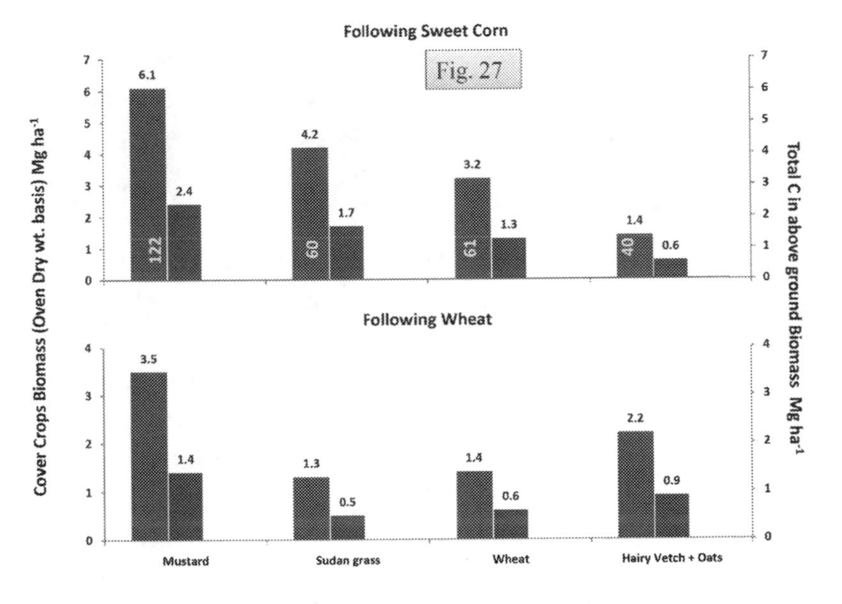

Production of total aboveground biomass and total carbon in the aboveground biomass for four cover crops in a potato-rotation system in the US-Pacific Northwest. The cover crops are planted immediately after harvesting corn or wheat in the Fall and incorporated with the soil in early spring during land preparation for planting the subsequent crop in a typical corn-potato-wheat rotation system in the US-Pacific Northwest. Mean of four years of field data.

This is an impressive option of removal of atmospheric CO_2 to mitigate its effects on climate change, by way of biomass production during the off season and depositing that C into soil that brings about additional soil quality benefits. Furthermore, this practice requires very little additional inputs and/or production efforts. Further studies with a range of cover crops as shown in the following figure

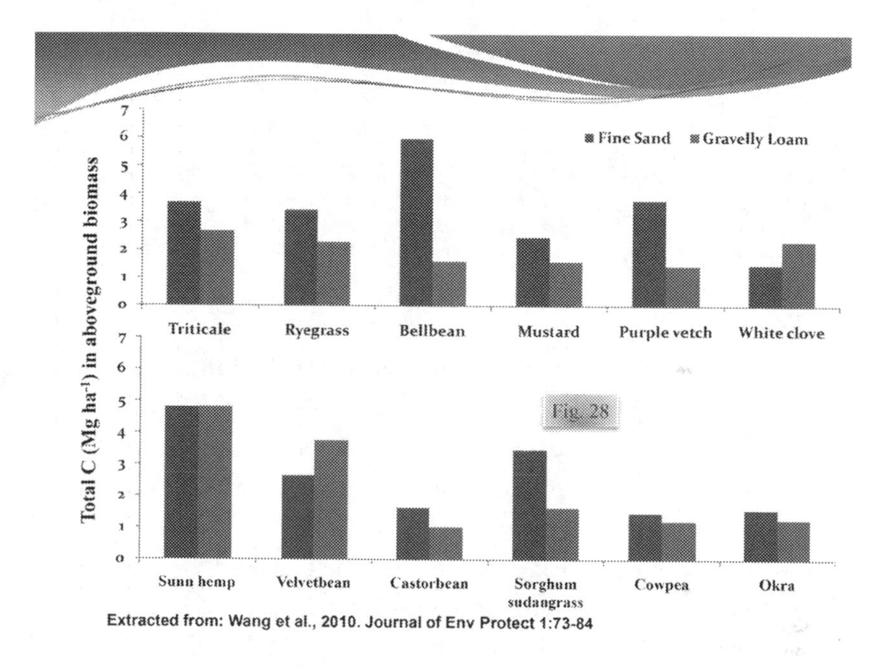

Total carbon in the aboveground biomass for different summer and winter cover crops grown on a Quincy fine sand from the US-Pacific Northwest, and a loamy soil from Florida. These soils are used for vegetable production under irrigated systems in the respective geographical region.

have demonstrated the accumulation of up to 5-6 tonnes per hectare of carbon in a sandy and loamy soil in warm, humid climate conditions.

12.3 Biofuels and Biochar Production

Increased concerns on potential negative effects of GHG emission from the use of fossil fuels on climate change as compared to that from biofuels has increased the emphasis on ambitious targets for increased production and use of biofuels. This trend has increased in the recent years due to a significant increase in cost of fossil fuels. The Energy Independence and Security Act (EISA) 2007 had mandate to produce 15 billion gallons per year (BGY) corn ethanol per year by 2015. That was equivalent to using 44 percent of total

US corn production at the 2007 production level. The target for cellulosic ethanol production is 16 BGY by 2022. The total quantity of ethanol production target (from corn grain and cellulosic conversion) accounts for a small fraction of (about 15 percent) of annual gasoline use in the US.

Water is an important somewhat limiting input for production of biomass needed as feedstock for production of ethanol. Depending on the crop evapotranspiration (ET) of different feedstocks used for ethanol, the water requirement (Liters of water; L_w) for production of liter of ethanol (L_e) is 500 to 4000 liters (Lw/Le). For corn the above data is 800 Lw/Le. Using an average of 16 miles per gallon (mpg) of ethanol, the use of corn ethanol would account for 50 gallons of water per miles driven (gwpm), i.e. 0.02 miles per gallon of water (mpgw). In addition, water requirement for processing a liter of ethanol from corn or sugarcane varies from 2 to 10 liters.

Therefore, although biofuel may be considered as an essential component for energy independence and security, the water requirement for biofuel production could be a major limitation, in addition to arable land and other crop production inputs. For example, the estimated nitrogen fertilizer requirement for feedstock production for 15 BGY of mandated biofuel production is about 2.2 million tonnes per year. This is roughly 16 percent of current nitrogen use for all crops production in the United States of America. As the need for biomass production increase to meet the biofuel production targets, there will be a greater demand for fertilizers which could increase the fertilizer price and in turn increase the prices for food and feed commodities. Furthermore, this could have some negative environmental impacts in terms of nutrient contamination of ground and surface water bodies, particularly under poor nutrient management practices, and expansion of production to vulnerable soils and ecosystems. These concerns warrant careful considerations to balance the priorities for food, feed and fuel productions.

The potential benefits of biofuel production by pyrolysis are described in the following figure.

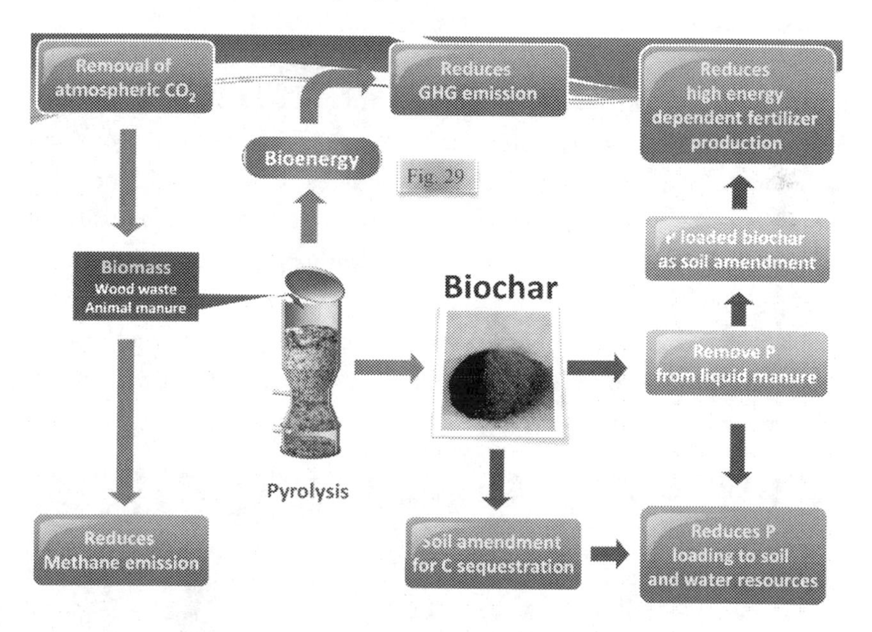

Schematic of feedstock used and byproduct of low temperature pyrolysis of biomass and animal manure for bioenergy production. The figure also shows reduction in trace gas emissions, and multiple uses of a byproduct of this technology, i.e. biochar.

In summary, using woody biomass or animal manure as feedstocks for pyrolysis reduces CO_2 and CH_4 emission into atmosphere. Utilization of bioenergy produced by pyrolysis, in turn, reduces GHG emissions into atmosphere. A byproduct of pyrolysis is biochar, which is rich in carbon and other nutrients that can be used as a soil amendment, thus contributes to carbon sequestration and other soil quality benefits. Biochar has very high surface charges, therefore, can be used to remove phosphorus from animal manure. This is a significant benefit to lower the phosphorus loading into soil when applying manure to agricultural soil as a primary source of nitrogen. Biochar loaded with phosphorus can be used as soil amendment as a source of plant available phosphorus, thereby provides an economical and environmentally friendly phosphorus source for agricultural soils.

Ethanol, a form of biofuel, is produced from corn grain or sugarcane. An alternative technology for production of ethanol is from biomass, generally referred to as 'cellulosic ethanol'. Woody biomass, or grain crop straw or stover can be used for this process. The most preferable biomass source crop used in this process should be low in lignin and must produce high amount of biomass per unit land per unit time, and with high water productivity. In this regard, switch grass is a good feedstock which produces high quantity of biomass per unit land area second year after the establishment onwards following a rather slow growth during the first year.

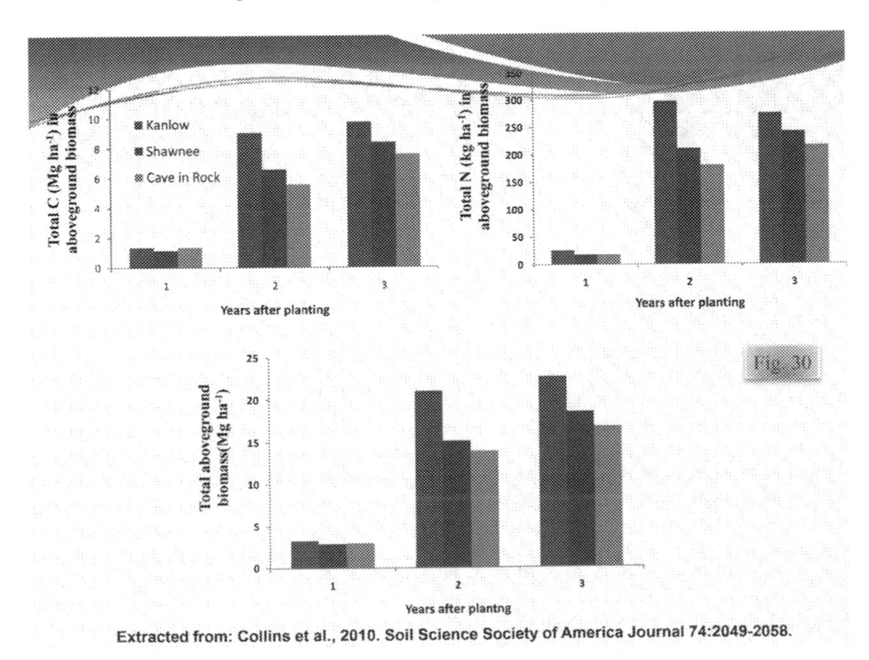

Total aboveground biomass and total carbon in aboveground biomass of three switch grass cultivars for the establishment year and subsequent two years in the US-Pacific Northwest under irrigation. Although the biomass productions of all three cultivars during the first year were rather low, the biomass increased substantially during the subsequent two years.

Total aboveground biomass in the US Pacific Northwest growing conditions during second and third year of establishment across three switch grass cultivars ranged from 15 to 23 tonnes per hectare, equivalent to 6 to 10 tonnes carbon per hectare. This region, for example Washington state, is conducive for production of high quantity of biomass per unit land area (Table 3) as evident from nearly 35 tonnes of biomass per hectare (4 year old stand) of a lowland cultivar (Kanlow), unlike 11 to 22 tonnes per hectare in other states including Texas, Alabama, Iowa, Nebraska. The respective biomass yields for an upland cultivar (Cave in Rock) are 23 and 6 to 18 tonnes per hectare.

Table 3. Switchgrass yields for several states in the United States of America (Unpublished data)

States	Switchgrass cultivar	
	Kanlow[1]	Cave in rock[2]
	Mg/ha (dry weight basis)	
Texas[3]	11.1	5.9
Upper South[3]	13.6	10.4
Alabama[3]	20.3	10.4
Iowa[3]	14.3	ND
Nebraska[3]	22.4	18.0
Washington[4]	34.6	23.2

ND= No data

[1] Lowland cultivar

[2] Upland Cultivar

[3] Biomass yield for a 6-yr old stand

[4] Biomass yield for a 4-yr old stand

The root biomass for three switch grass cultivars in Washington State, third year after establishment, varied from 3400 to 4400 kg per hectare in the top 15 cm depth soil and from 1400 to 2600 kg per hectare in the 15 - 30 cm depth.

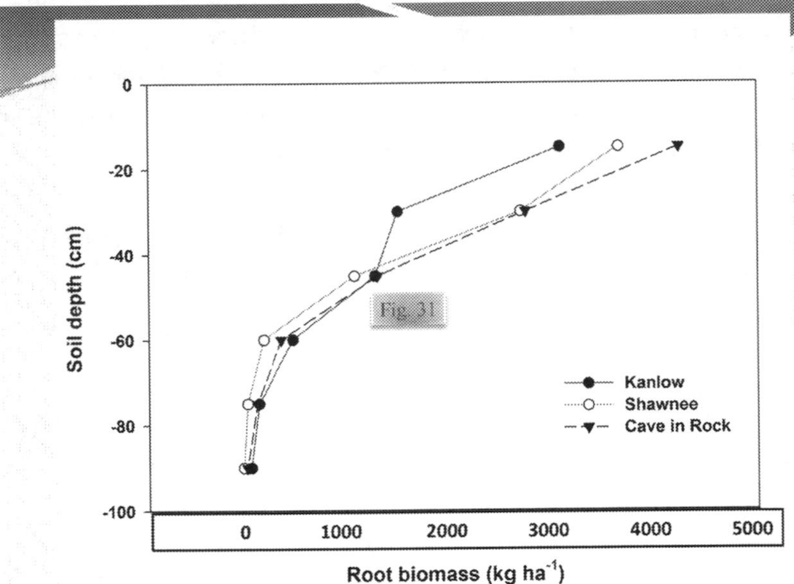

Root biomass of three switchgrass cultivars at various depths in the soil profile in irrigated US-Pacific Northwest production conditions

This is an important consideration to ensure carbon sequestration and improved soil quality and health when using the arable land for biomass production to support biofuel production targets.

The energy and carbon balance from different bioenergy feedstocks are shown in Table 4. The data clearly demonstrate distinct benefits of using switch grass as a feedstock in comparison to other feedstocks such as corn stover or wheat straw or grain corn. Certainly, it is highly recommended to consider these benefits in choice of suitable feedstock for bioenergy production that would be sustainable and will have least impact on the food production.

Table 4. Estimated energy flow in producing ethanol from corn grain, corn stover, wheat straw and switch grass under irrigation (Extracted from: McLaughlin and Walsh, 1998)

Process	Corn (Grain)	Corn (Stover)	Wheat (straw)	Switchgrass
GJ ha^{-1} yr^{-1}				
Crop Production[a]	37.4	37.4	28.3	30.3
Biomass energy[b]	257.8	315.8	240.7	521.6
Evergy ratio[c]	7.0	8.4	7.8	17.2
Ethanol Production[d]	54.2	10.3	10.3	10.3
Energy in ethanol[e]	126.4	75.4	52.3	221.8
Total energy ratio[f]	1.37	1.58	1.35	5.46
Net energy gain[g]	37%	58%	35%	446%

[a]Crop production based on budget data from USDA, 2002. [b]Biomass energy based on crop yield data. [c]Biomass energy/production energy. [d]Processing distribution energy, credits for co-products of corn and combustion of lignin from switchgrass. [e]Ethanol yields, 23.3 MJL^{-1}, assumed 60% removal of corn stover and wheat straw. [f]Total output energy/input energy.

12.4 Animal Manure Management - Anaerobic Digester:

Emission of CH_4 from animal manure is a major source of GHG. In most cases animal manure is stored in lagoon, until disposed as soil amendment for forage crops. This is somewhat an inefficient process due to limited land area available in close proximity of manure generation facility. This limitation further contributes to heavy loading of liquid manure on a unit land area in an effort to minimize the cost of disposing large quantity of animal manure. Such a practice leads to overloading of nutrients, mainly phosphorus and nitrogen, and in turn, potential for nutrients runoff and/or leaching resulting in surface and/or groundwater contamination. The ratio between P to N in the manure is quite larger than that of nutrient uptake by the crops. As a result, when manure is applied based on N requirement of the crop, there will be a large reserve of P in the soil. This trend continues during multiple years of manure application. Case in point is excessive P accumulation in vegetable greenhouses in China, to the tune of close to 180 to 200 mg/kg soil of available P in the top soil (measured by Olson extraction). This was the result of heavy loading of manure to greenhouse vegetable production, which resulted in annual P loading of >570 kg/ha from manure. Removal of P from the soil as per P requirement of the vegetables is <50 kg/ha, thus contributed to a large P build up in soils.

This impacts sustainability of animal management thus needs careful reevaluation to mitigate the perceived conflict between the agriculture/food systems vs. sustainability/environmental quality. The manure storage lagoon and manure spray fields both contribute to methane emission. Emission of methane into atmosphere is not only a loss of energy but also has negative effects on climate change. These problems can be overcome by adaptation of anaerobic digester technology to capture methane from manure and use it as an energy source.

During anaerobic digestion of manure, methane is captured and stored as a source of energy generation. The byproducts of this process include anaerobic digester effluent (ADE) and fiber (ADF) both can be utilized for soil application as sources of plant available nutrients and for carbon sequestration. Studies have shown low levels of carbon dioxide and methane emissions with land application of ADE or ADF as compared to those from soil with application of liquid manure (LM) directly from lagoon without anaerobic digestion. Therefore, adaptation of anaerobic digester provides a source of sustainable, eco-friendly bio energy production, in addition to reducing GHG emissions into atmosphere from animal manure.

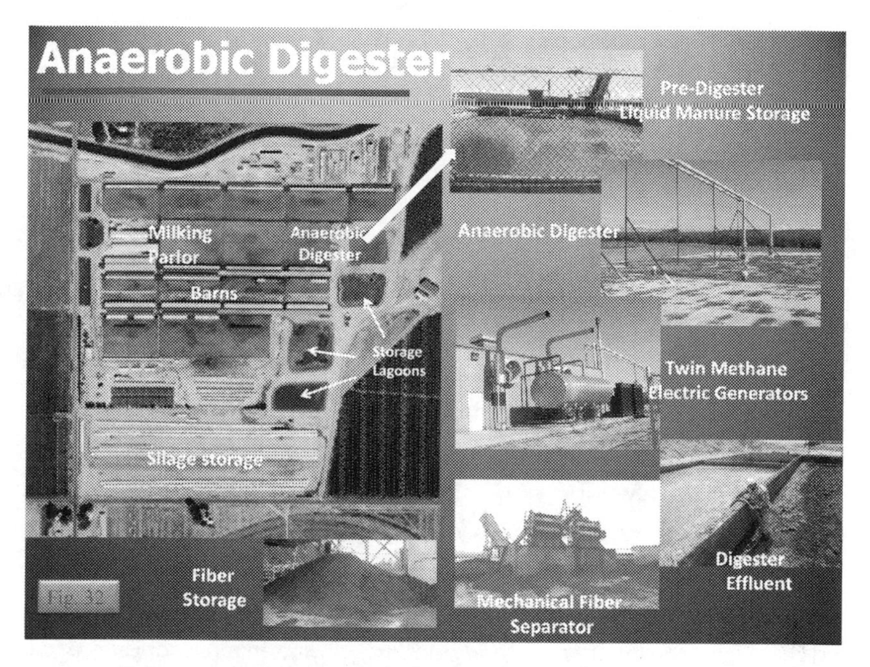

Schematic representation of anaerobic digestion of animal manure used for energy generation in a large dairy farm

12.5 Nutrient and Water Management:

Nitrogen fertilizer applied to soil for crop production is a source of nitrous oxide emission, which is also a greenhouse gas that can impact climate variability. Therefore, management options to improve nitrogen uptake efficiency and mitigate conditions that favor nitrous oxide formation in the soil are important to the mitigate nitrous oxide emission to atmosphere from agriculture. Application of encapsulated calcium carbide (ECC), which is a form of nitrification inhibitor, has shown to reduce the emission of nitrous oxide from both corn and wheat fields (Table 5). Likewise recent studies have also provided evidence of reduced emission of nitrous oxide, methane, and carbon dioxide under lowland rice production conditions with application of ECC.

Table 5. Effect of Encapsulated Calcium Carbide (ECC) on N_2O emissions from irrigated wheat and maize (Mosier et al., 1996)

Treatment	Wheat	Maize
	($g\ N_2O$-N/ha)	
Urea	930	1650
Urea + ECC	510	440
No N control	440	110

Flooded soils under lowland rice production contribute to methane emission. Most of the rice production in predominantly rice based diet regions in the highly populated Asian countries is in flooded soils. Further studies need to be conducted on long term evaluation of rice yield and quality differences between continually flooded vs. intermittent aeration water management conditions. However, the methane emission was reduced under intermittent flooding and aeration conditions as compared to that with continuous flooded water management.

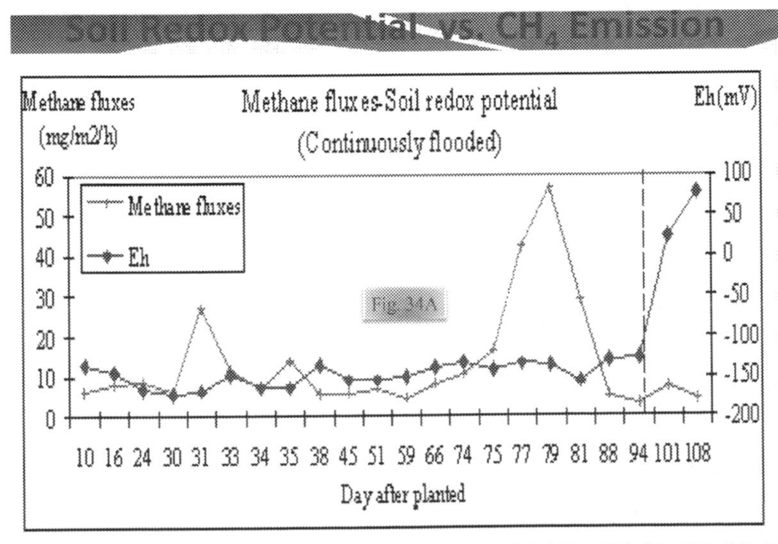

Towprayoon, 2004, Greenhouse Gas Mitigation Options from Rice Field. Presented at In-session workshop on Climate Change Mitigation 19 Bonn 2004, Maritim Hotel, Bonn

Effects of several intermittent short duration aeration of flooded soil, on redox potential and methane emission.

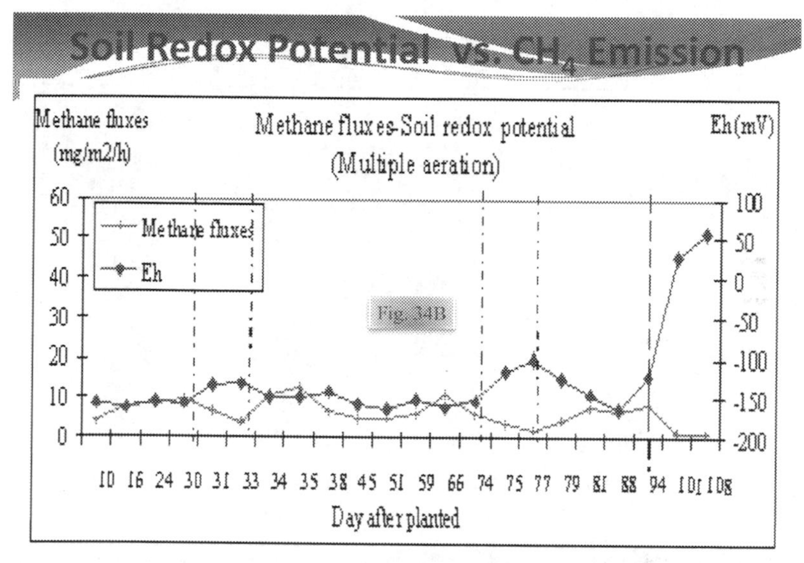

Towprayoon, 2004, Greenhouse Gas Mitigation Options from Rice Field. Presented at In-session workshop on Climate Change Mitigation 19 Bonn 2004, Maritim Hotel, Bonn

Methane emission is linked to redox potential (Eh) of flooded soils. This is an index of anaerobic status of the soil due to depletion of oxygen in the soil pores during flooding. The Eh measurements in the above study remained low most of the growing season with continuous flooding. On the contrary short duration aerations increased the Eh which, in turn contributed to reduced methane emissions.

Studies conducted in Rothamsted Experiment Station, United Kingdom, have shown that winter wheat yield increased from 2.1 to 9.1 tonnes per hectare with an increase in nitrogen application from 0 to 288 kg per hectare.

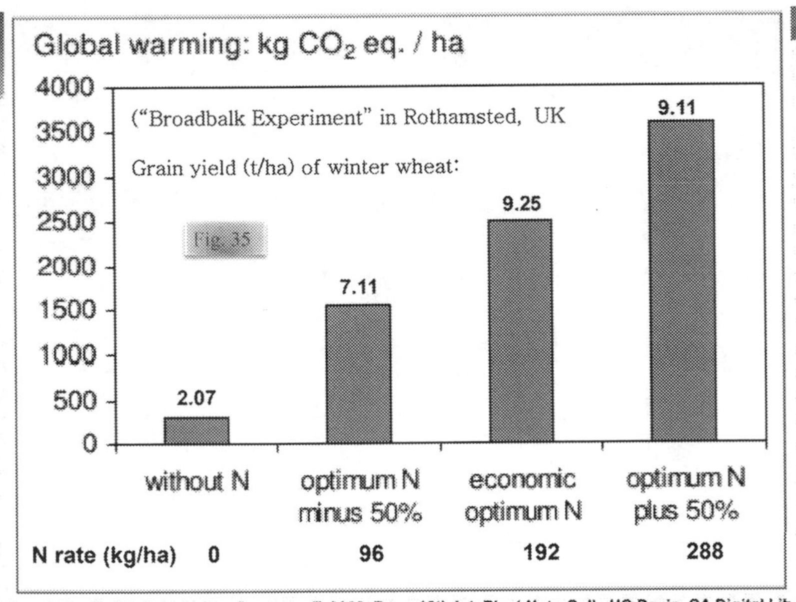

Estimated global warming potential under different N management for winter wheat production and grain yield response.

The economic optimum nitrogen rate was 192 kg per hectare. The global warming potential, expressed as kg carbon dioxide equivalent per hectare increased proportionally with an increase in nitrogen rate. The Fig. 35 clearly shows that application of nitrogen rate above the economic optimum nitrogen rate of 192 kg per hectare not only failed to increase the grain yield but also contributed to an increased level of global warming potential. However, to produce a given amount of yield equivalent to that with economic optimum nitrogen rate, increased acreage needs to be brought under cultivation at lower nitrogen rate. This in turn increases global warming potential due to carbon dioxide release from increased land area put to cultivation to compensate for lower grain yield for unit land area under no nitrogen or at 50% of economic optimum nitrogen rate treatments.

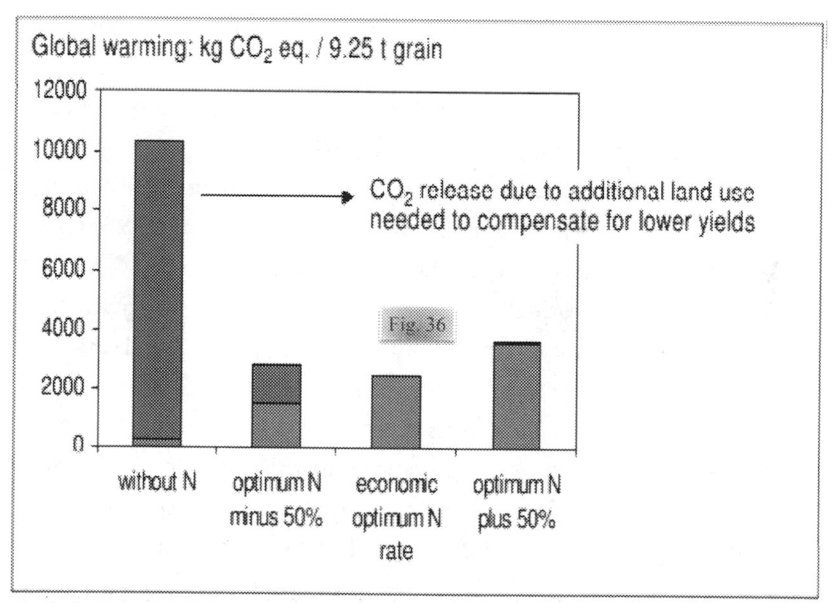

Reproduced with Permission: Brentrup, F. 2009. Proc. 16th Int. Plant Nutr. Coll. UC Davis, CA Digital Lib.

Global warming potential across different N rate for winter wheat production with target production of grain yield equivalent to that with economic optimum N rate application.

Thus, Fig. 36 shows that global warming potential is lower with application of economic optimum nitrogen rate if the target is to produce the similar total grain yield at low nitrogen rate per unit land area.

Summary of Agricultural Practices to Mitigate GHG Emission

Although agricultural practices contribute to GHG emissions, economically sustainable and technically feasible strategies are available to mitigate GHG emissions and enhance carbon dioxide removal from the atmosphere by high production of biomass and deposit the carbon in the biomass into the soil, i.e. carbon sequestration. The latter approach has multiple benefits, i.e. reduce soil erosion, increase soil quality and nutrient retention, and enhance soil productivity. Reduced tillage and growing cover crops have demonstrated decreased emission of GHG and transfer of atmospheric carbon dioxide into soil to enhance carbon sequestration and associated soil quality benefits. With recent years greater emphasis on energy security and targets for increased production of biofuels requires large agricultural land area to be diverted for production of biofuel feedstocks. Switch grass is an efficient feedstock with high biomass production that can be grown in marginal land with reduced amount of inputs, thus offer increased utilization of atmospheric carbon dioxide and carbon sequestration in the soil.

Pyrolysis of woody biomass and animal manure produces bioenergy and a byproduct rich in carbon, i.e. biochar that is an excellent source of soil amendment for carbon sequestration. Furthermore, our studies have shown that biochar can be used to remove phosphorus from liquid animal manure, thus minimize the risk of heavy phosphorus loading into soil by application of liquid animal manure containing low levels of phosphorus hence can mitigate surface water contamination of phosphorus. Biochar enriched with phosphorus is a good source of plant available phosphorus in addition to its soil quality benefits when this is amended to agricultural soils. Biochar amendment to soils also reduces nitrous oxide emission.

Liquid animal manure stored in lagoon and/or applied to agricultural soils at high rate (as a disposal mechanism) contributes to methane emission. Methane can be captured as a source of energy in anaerobic digester. The effluent and fiber from this process are good nutrient sources for soil amendment which contribute to lower methane and nitrous oxide emission unlike soil application of liquid animal manure without processing through anaerobic digester.

Improved nutrient and water management for crop production also contribute to reduce GHG emissions. Application of encapsulated calcium carbide (ECC; urease inhibitor) with urea has demonstrated to decrease nitrous oxide emission from wheat, maize, and rice fields as compared to that from the respective crop fields with application of urea without ECC. Continuous flooding of lowland rice causes anaerobic conditions resulting in methane emissions. Short duration aerations of flooded soils have shown to decrease redox potential of the soil and, in turn, decreased methane emissions. The strategies discussed above if adapted on a large scale will have a significant role in mitigation of GHG emissions from agricultural practices to overcome the negative effects on climate variability.

13

Changes in Dairy Farm

Modernization and mechanization have made all facets of agriculture more efficient and less dependent on labor force. This is also evident in commercial dairy industry. Mechanization has made it possible to increase the number of animals that can be managed in a dairy plant.

As a result the number of US dairy plants decreased from 3700 in 1970 to almost 1300 by 2010, i.e. almost 65% reduction in 40 years.

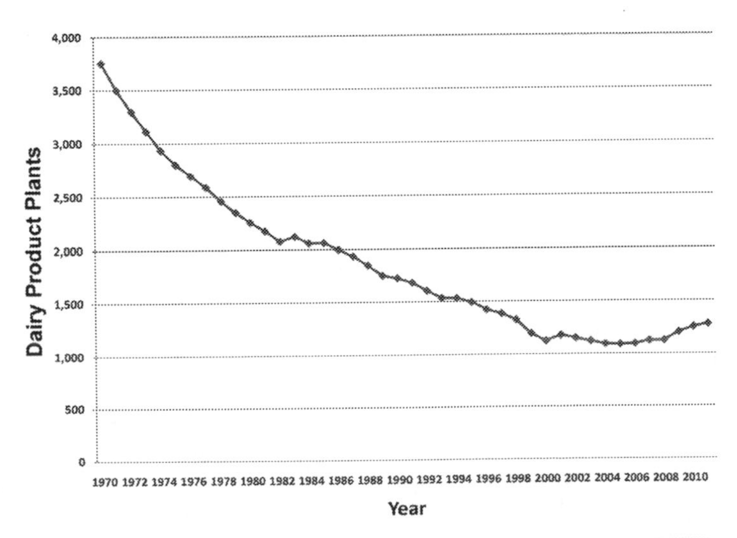

http://quickstats.nass.usda.gov

Fig. 38

Decline in number of dairy plants in the United States of America from 1970 to 2010. The trend is a reflection of increased mechanization that may have contributed to large scale dairy farms.

The number of dairy cooperatives decreased from 2270 in 1935 to 241 in 1995 with a corresponding decrease in membership in cooperatives from 720,000 to 117,313.

Table 8—U.S. dairy cooperative statistics, 1935-95, selected years

Year	Cooperatives	Members	Milk marketed to plants and handlers by cooperatives[1]	Business volume	Cooperative share of total milk delivered to plants and handlers
	Number		*Million pounds*	*Million dollars*	*Percent*
1935/36	2,270	720,000	31,058	520	48
1943/44	2,286	702,000	NA	1,203	NA
1956/57	1,746	777,240	53,038	2,764	59
1964	1,244	561,085	76,743	3,524	67
1973	592	281,065	83,227	6,102	76
1980	435	163,549	95,634	13,666	77
1987	296	120,603	105,798	16,548	76
1992	265	110,440	122,622	20,239	82
1993	258	122,396	127,090	20,510	86
1994	247	124,666	129,780	21,503	86
1995	241	117,313	NA	21,784	NA

NA = Not available.
[1]ERS estimates for 1993 and 1994

The Structure of Dairy Markets: Past, Present, Future/AER-757 Economic Research Service, USDA

Fig. 39

Changes in the role of dairy cooperative in the United States of America from 1935 to 1995.

For the same period amount of milk marketed increased from 31,000 to almost 130,000 million pounds, with a business volume of 520 to 22,000 million dollars. Nearly 86% of milk delivered to plants and handlers came from cooperatives in 1995, unlike 48% in 1935. Therefore, clearly the trend was an increase in large scale commercialization and mechanization of dairy operations.

14

Various Components of Food Price

With increasing specialization of production and market system of food products, there are number of players at each step starting from growing crop until the product is ready to be available in the market for consumers. Of every dollar consumer pays for food product, in the United States of America, only 10 percent can be traced to the cost of producing the crop, i.e. farm and agribusiness.

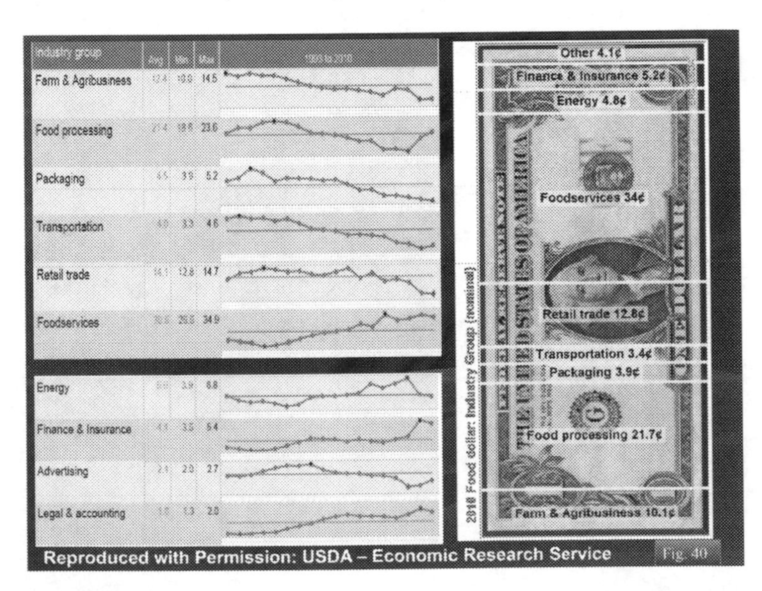

Analyses of various components of consumer price of food dollar. Notice that the costs associated with farm and agribusiness combined accounts only 10 percent of the retail food price consumer pays. The figure also illustrates the changes in proportions of various components of food cost during 1993 through 2010.

The major component of food price, i.e. almost 56% is associated to food services and processing industries. Retail trade contributes to 13%. Costs associated with energy, finance, and insurance account each at 5%. Packing cost is 4%, and transportation is 3.5%. Most consumers are not aware of the fact that only 10% of food price is associated with cost of producing the crops. Therefore, the recent hikes in food price is not due to an increase in cost of production, instead is likely due to an increase in other component costs shown in Fig. 40.

An evaluation of changes in various components of food cost during 1993 to 2010 revealed that the costs of energy, food services, finance/insurance, and legal costs have increased. All other components costs have decreased during the same period. This is clear indication that inflation adjusted fraction of food dollar going to farms and agribusiness, i.e. for high risk and economically uncertain food production sector, is extremely small. Only way farmers are able to cope this is by increasing production efficiency by adaptation of new technologies and best management practices. It is important to underscore the role of research in improving efficacy of agricultural production management. Therefore, continued investment in agricultural research is critical to support a vibrant agricultural industry able to continue to provide abundance of high quality safe food at low price despite continuing to face various production challenges and emerging problems, including need to produce more to feed the increasing population, uncertainty due to climate variability, and emerging pests and diseases. Most economic analyses confirm 20 – fold greater returns for a dollar invested in agricultural research.

Indeed modern day farmers in the developed countries have to be very creative and innovative to make a sound balance between external pressures from consumers to produce high quality and safe food in abundance at low price and from environmental and regulatory sectors to maintain sustainability to minimize potential negative impact on the environment. The impact of the above external factors and inherently high risk and relatively low net return of farming industry make the farming somewhat a risky venture, therefore, small farms go out of business. To some extent the net result of multiple risks and uncertainty on US farming in part contributed to dramatic shrinkage in fraction of the US population engaged in farming enterprise.

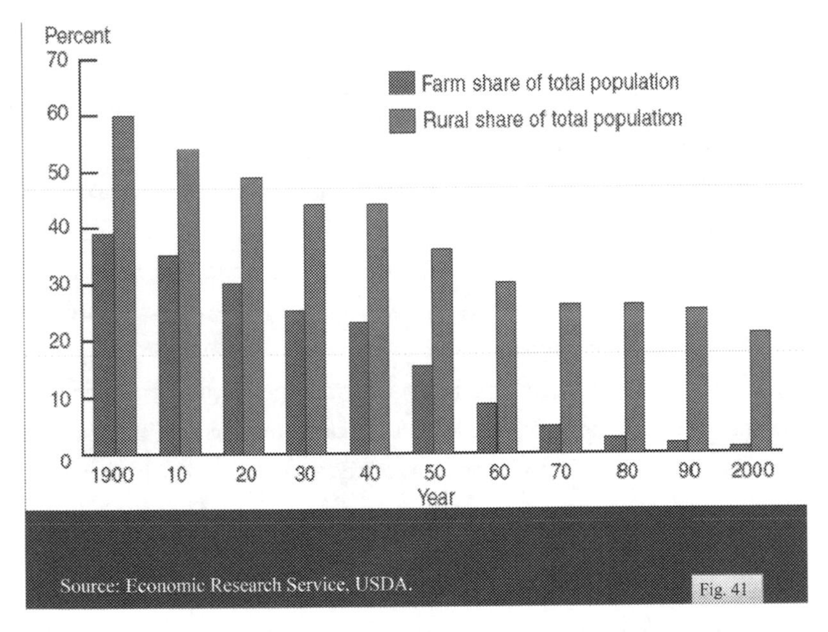

Changes in percent of US farm or rural population from 1900 to 2000. Notice the steep decline in farm population as compared to the rural population.

In the early 1900, the US population directly involved in farming was close to 40% and decreased to < 2 % by 2000. This decline was substantially greater than that of rural share of US population, i.e. a decline from 60 to 25% for the similar time period.

Gross domestic product (GDP) is the total wealth of a country from all sources. In the early 1900's the contribution of agriculture to the nations GDP was up to 7.5% with up to 40% of the workforce employed in agriculture. By 2000, only 0.7% of the nation's GDP originated from agriculture with only <2% of workforce employed in agriculture. Despite the rapid increase in agricultural production and economic development during this period, the industrialization and overwhelming increase in other economic sectors in the US explain, in part, shrinkage in contribution of agricultural economy to nations total GDP. Similarly modernization in agriculture and increased use of machinery for handling large scale agricultural practices contribute to drastic reduction in percent of total workforce employed in agriculture from 1990 to 2000. Overall, the US modern agriculture is highly efficient and specialized with increased production per unit land and resources, less labor intensive, high input based, and increasingly low net return except some years of high commodity prices, which is very cyclic. Furthermore, market globalization had both positive and negative impacts on the agricultural industry. This has created new and promising markets for the US farm products around the world.

15

Biofuel Production: Future Challenges

Global biofuel production increased from 16 to >100 billion liters during 2000 to 2010.

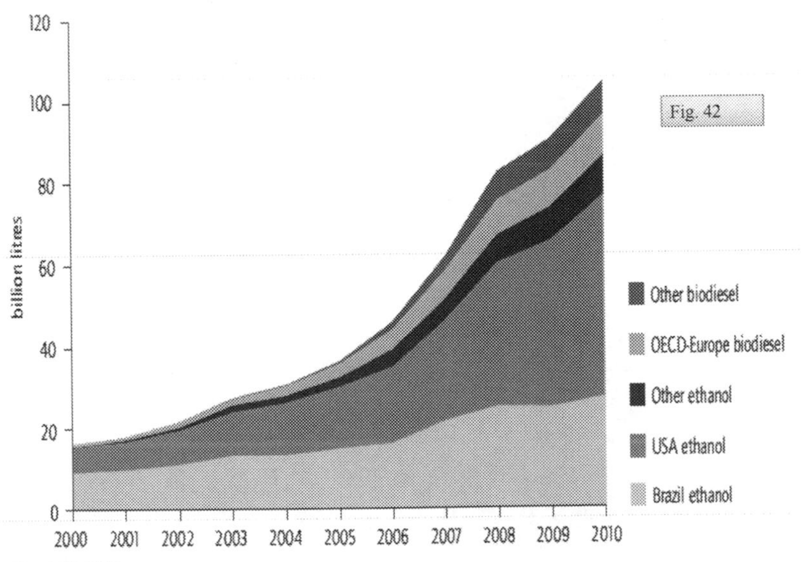

Source: IEA, 2010a.

Reproduced from: http://www.iea.org/publications/freepublications/publication/biofuels_roadmap_web.pdf

Growth in ethanol and biodiesel production during 2000 to 2010. Notice the steep increase in US ethanol production post 2005.

Brazil was undeniably a world leader in biofuel production up until 2005. Subsequently, US ethanol production increased sharply in quantities far exceeding that of Brazilian production. This was driven by the US policies and priorities towards increased emphasis on energy independence due to political unrest in the major oil producing countries in the Middle East. These unrests have contributed to a large increase in fuel price until 2015. Since then the oil price began to slide rather rapidly as a combination of several factors, which appeared to create an indication of glut in the market. Detailed discussion of the factors contributed to decline in world oil price is not within the scope of this book. Furthermore increase in disposable incomes of the emerging economy countries have increased the demand for fuel which also contributed to increasing fuel price, until the 2015. By 2010, the US ethanol production was almost two fold that of Brazil. The main difference between these two countries is that the Brazilian ethanol production is sugarcane based unlike corn based ethanol in the US. Biodiesel is a small component of the total biofuel production.

The projection for 2050 on the energy use for transportation shows that diesel and gasoline combined will continue to be the dominant energy source (i.e. 36%).

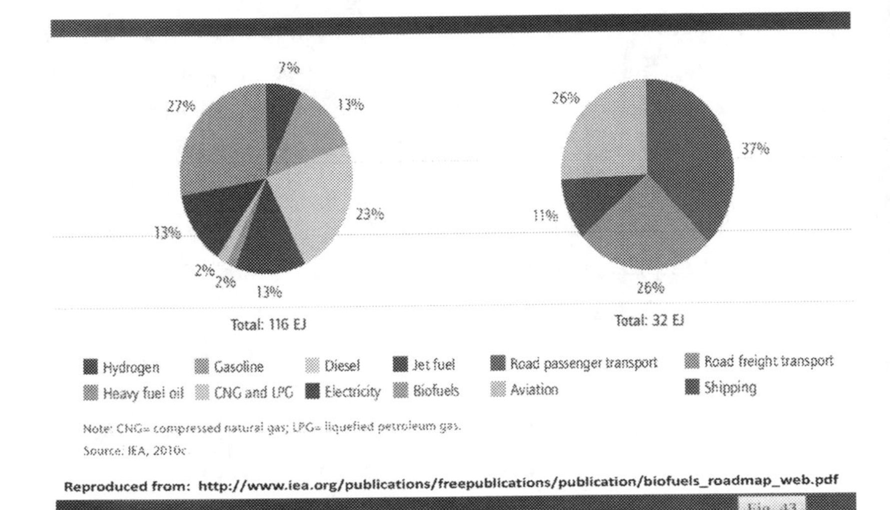

Percent of different energy sources including biofuels, and percent of energy used for different transportation systems. Notice that nearly 50 percent of total energy sources are accountable to gasoline, diesel, and jet fuel. Road transportation for passengers and freights accounts 63 percent of total energy for all transportation systems.

Biofuel is expected to become a dominant player as source of energy for all transportation by 2050, i.e. 27%. Electricity is expected to contribute 13% of all energy use in transportation. Tremendous expansion in aviation as a result of globalization is evident from aviation fuel accounting for 13% of all transportation fuel. The projections for 2050 on the share of biofuels for different transportation sectors reveal 37, 26, 26, and 11% of biofuel used for passenger transportation, road freight, aviation, and shipping, respectively. Progress in biofuel production is however impacted by global fuel price, availability of other sources of fossil fuel such as fracking, oil shale etc, and commitments and policies of industrialized countries to reduce greenhouse gas emissions.

The efficiency of ethanol production from different feedstock is judged based on the quantity of ethanol produced per unit land area.

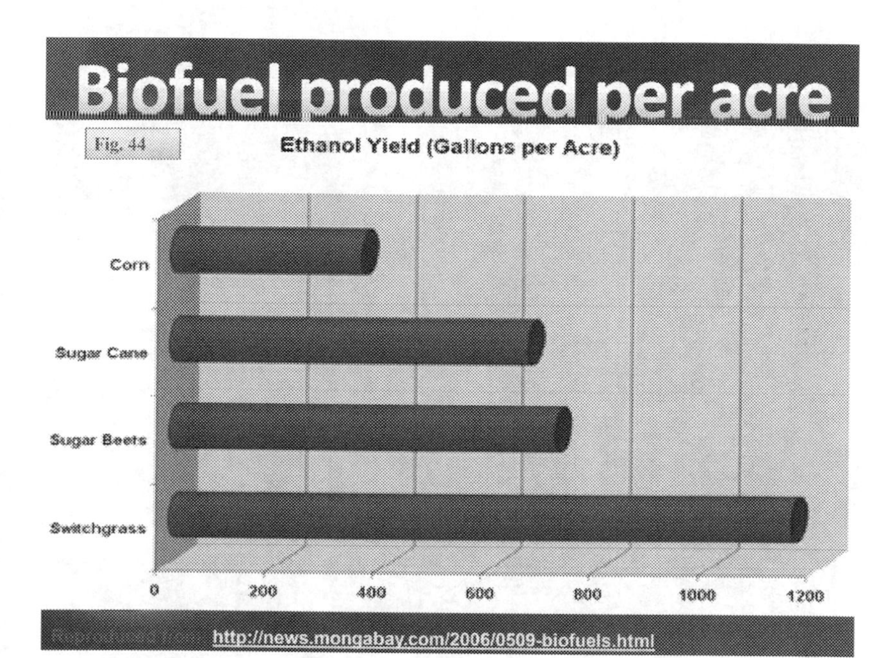

Ethanol yields (gallons per acre) from different feedstock. Ethanol yield from switch grass is four folds greater than that from corn.

This evaluation is important because as the demand for biofuel increase the need for biofuel production increase which requires more land area needed for feedstock production. The availability of land area and production inputs, mainly water, will then become the primary limiting factor for supply of increased demand for feedstock to cope with increased production of biofuel.

On a per acre basis, ethanol output is < 400 gallons for corn, 600-700 gallons for sugarcane and sugar beets, and >1000 gallons for switch grass. Brazil has been very proactive in production of ethanol well before this became a priority in the US. Sugarcane is the main feedstock for ethanol production in Brazil, unlike corn in the US. In the early 1980's, Brazil reported almost 3 million acres of sugarcane being used for ethanol production, while US ethanol production was quite insignificant i.e. < 1 million acres of corn.

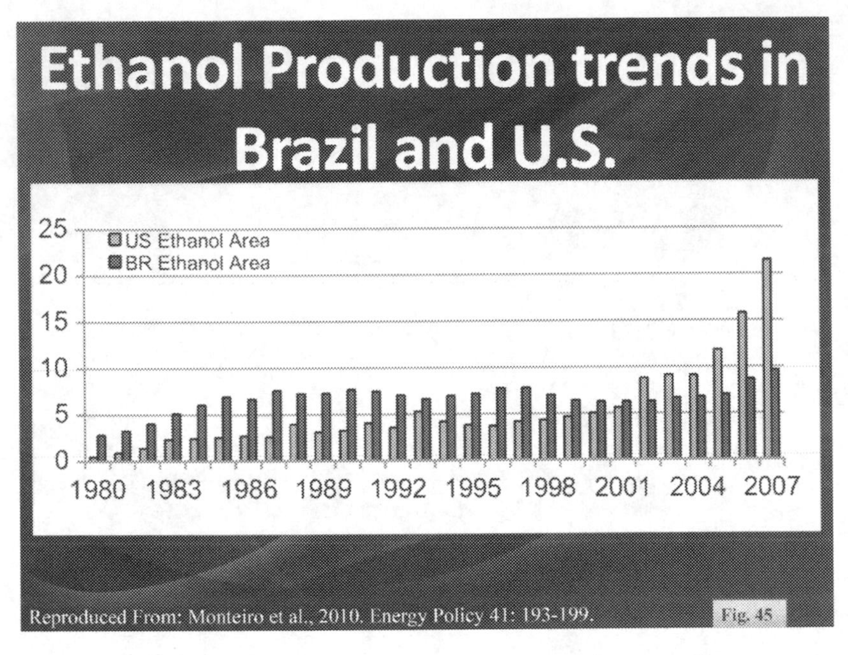

Changes in land area used for ethanol production in the United States of America and Brazil from 1980 to 2009.

The acreage devoted for feedstock production increased rapidly in Brazil and attained the peak of about 7 million acres by 1987 and remained at this level until 2005. This acreage further increased to 9-10 million acres in 2006 and 2007. In US on the other hand, the corn (for ethanol) acreage increased steadily until 2001, and by 2002 the corn acreage in the US (for ethanol) was greater than the sugarcane acreage in Brazil. From 2001 to 2007 there was a rapid increase in corn acreage in US for ethanol production which was 22 million acres by 2007. This was greater than 2 –fold of sugarcane acreage in Brazil. However the ethanol production efficiency is almost 1.5- fold greater for sugarcane as compared to that for corn.

Using the 2007 data, production of ethanol in Brazil was approximately, 10 X 10^6 acres of sugarcane X 610 gallon ethanol/ acre= 61 x10^8 gallon ethanol per year. In contrast the ethanol production in US in 2007 was approximately using 22x10^6 acres of corn x 350 gallon ethanol/acre = 77 $\times 10^8$ gallon ethanol per year. Therefore, for a 2-fold greater land area devoted for feedstock production for ethanol in the US as compared to that in Brazil, the total ethanol production was only 1.25 – fold greater in US as compared to that in Brazil. This clearly exemplifies that on land area basis sugarcane is a preferred feedstock as compared to corn for production of ethanol. However it is necessary to consider the other economic factors i.e. cost of production of different feedstocks on unit land area basis and also land suitability and availability of water for production of different feedstocks. The water use for sugarcane is certainly greater than that for corn.

Based on the ethanol production in US and Brazil, the latter maintained 80% of the total ethanol production until 1991,

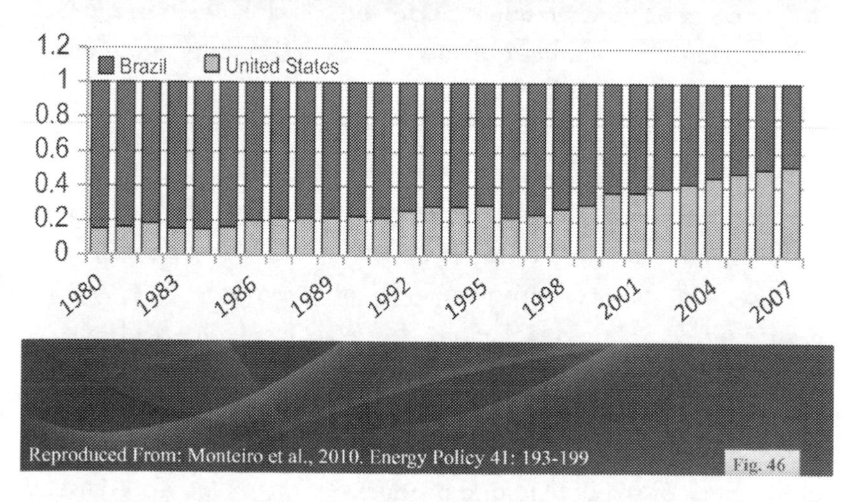

Historical trend in proportion of Ethanol from Brazil and that from United States of America during 1980 to 2007.

with remaining 20% production in the US. Since the early 1990's, the ethanol production in the US increased at a slow phase till 2000 and then at a rapid phase post 2000. By 2007, the proportion of US vs. Brazil ethanol production was about 52:48. However, as described previously this was attained at the expense of almost 2- fold greater acreage utilized for corn production in the US as compared to that for sugarcane in Brazil for ethanol. This is due to the difference in ethanol production efficiency of corn vs. sugarcane.

Recently, there has been greater emphasis on ethanol production by cellulosic conversion using any biomass such as corn stalk, wheat stover, forest biomass, and switch grass. Increased research on this technology when perfected, will give considerable advantages in further boosting ethanol production due to the diversity of biomass feedstocks available. A lot of feedstocks have no real economic value at the current market, i.e. biomass of various crops, which are generally plowed into the soil. This practice has greater significance in enhancing soil chemical, physical, and biological properties, i.e. soil health improvements, and enhances carbon sequestration. Therefore, massive utilization of biomass for biofuel production in the future when this technique is perfected will deprive its use for soil incorporation which can negatively impact soil quality and carbon sequestration.

This is an example of conflicting outcome of creating a problem while trying to solve one problem. In an effort to gain energy independence and reduced reliance on fossil fuel, there has been greater emphasis on increased production of biofuels domestically. This is a novel goal to support energy independence and also use green energy with less pollution. However, the indirect effects of this priority are: (i) competition between food crops and biofuel feedstock crops for available and rather limited resources, i.e. land and water; (ii) potential increase in food price; (iii) potential deterioration of soil quality and productivity due to increased use of crop residues for ethanol production which deprives the availability of biomass for soil incorporation; (iv) negative effects on natural and forest ecosystems by potential increase in use of biomass for biofuel production; (v) potential expanded use of land for biofuel feedstock production, which is currently designated for special programs aimed for improvement of conservation and soil quality, i.e. fallow land, land under conservation reserve program (CRP), and marginal land with limited productivity. Continued use of land under CRP may cause further degradation of land resulting in irreversible environmental damages. We need intensive studies with the above aspects to evaluate agronomic, ecosystem, socio- economic aspects of land use for biofuel feedstock production so that unbiased results are available for the policy makers to make the right decision to maintain sustainability and address the energy independence. We have to evaluate the impact of the above negative effects against the potential benefits of using biofuel in terms of reduced greenhouse gas emissions which has an impact on the climate variability.

16

Regulatory Steps in Fertilizer Management

State and Federal governments have an authority and obligation to maintain clean water supply for the current and future citizens. To this end the government must act to minimize potential contamination of ground and surface water bodies by any contaminants, i.e. industrial, municipal or agricultural. The agricultural contaminants to water bodies include: (i) chemicals used as pesticides or herbicides; (ii) nutrients from fertilizers especially nitrate form of nitrogen and phosphorus (iii) nutrients and pathogens derived from animal manure i.e. nitrate, phosphorus and coliform.

There are two contrasting approaches state can act to minimize water contamination, i.e. regulatory or proactive voluntary. The former is rather short sighted and often faces high headwind from the farming community with a lot of complications, lost trust, and resentments. The latter is generally preferred approach. The farmers are empowered to take ownership for being good stewards of land. This approach may include some short term and long term incentives. This is often done by developing best management practices (BMP) based on long term research by land grant universities. The success of this program largely depends on the close cooperation between the researcher, farmers, agribusiness, and regulatory agencies in all steps in the process, i.e. planning and conducting research, review of results and developing recommendations. It is preferable to conduct research on farmers' field, and organize field days for other farmers to see the research in progress and present research update that can eventually contribute to developing Best Management Practices (BMPs). This kind of research is often high cost due to need to make various measurements including tracking the nutrient and/or chemicals in the soil, plant and ground water. In addition, it is also necessary to evaluate crop yield and product quality responses over a long term evaluation to ensure adequate reproducibility of the responses. The current funding of most land grant universities is inadequate to support this extensive long term field studies, due to limited budget and often funding priorities are primarily driven by other high impact research needs and emerging problems.

Therefore, there is a need to identify different sources of research funds for this kind of long term, high cost applied research. The following is an example to describe the process of developing nitrogen BMP.

The public water supplies in urban areas by the city and municipal water authorities are highly regulated to maintain safe drinking water standards. These water supplies are routinely tested for various contaminants to ensure that the levels don't exceed the recommended allowable limits. This system of testing is not extended to private wells in rural areas which are outside the boundaries of serving areas by the public water supply agencies. Public Health Service in most states monitor the quality of drinking water from private wells in rural areas. This is done on random sampling basis with very limited data on the well construction details and possible leaks from the well casing.

In the early 1900's the drinking water monitoring of rural area limited number of randomly sampled private wells revealed a consistent pattern of increase in nitrate concentration in increased proportion of wells sampled in Florida (FL) over the years. This prompted the FL legislature to pass nitrate BMP bill which authorized the FL Department of Agriculture of Consumer Services (FL DACS) and Department of Environmental Protection (FL DEP) to address this problem. In order to accomplish this goal, FL DACS and FL DEP worked with land grant University scientists and extension specialists to develop crop specific N BMPs, i.e. 'practices or combination of practices determined by research or field testing in representative sites to be the most effective and practicable methods of fertilization designed to meet nitrate groundwater quality standards, including economic and technological consideration'.

The legislature was very considerate of the fact that the final BMP must meet economic and technological considerations. This was necessary to ensure that the practices developed can be readily adaptable by the grower without the need for additional investment to accommodate the new technology and also that it should not drastically increase the cost of input which would translate into reduced net returns. In order to fund this research, the bill authorized FL DACS to assess and collect an additional tax on all N fertilizers sold in the state of FL. This funding was to be utilized on major needs, i.e. on crops which appeared to show greater impact on the potential for nitrate contamination of groundwater as evident from the water quality monitoring data. So initially the priority was given to developing N BMP for citrus, fern, and vegetables. The research funding was administrated by a panel which included mostly growers, agribusiness leaders, and researchers for reviewing and approving the project proposals. FL DACS was coordinating this granting process. This was rather a smart decision by FL DACS to empower the end users in decision making process. Open field days and workshops for presentation of progress reports where conducted throughout the project phase to provide an opportunity for producers and other researchers to review the results and make constructive comments and suggestions.

At the conclusion of the long term (7 years) research, the results were reviewed by an independent panel of statewide experts from the land grant university (plant nutrient recommendation committee). Subsequently, FL DACS used the project results for nitrate BMP rule making specific for the crops. The draft rule was presented to the grower organization for comments, and adjustments were made to accommodate the growers and agribusiness comments, on case by case basis. Upon completion and finalization of the Nitrate BMP rules, FL DACS then provided an opportunity for growers voluntarily enroll to follow the Nitrate BMP recommendation that evolved from a thorough process as described above. Those growers who voluntarily adapted the BMP were given waiver of liability for potential water quality impact, as an incentive. The upshot of this program, for example for citrus was that the parallel large scale farmer field demonstration of application of the BMP revealed that a combination of optimal source, timing and frequency of N application contributed to: (i) reduced the rate of total N required for high yields as compared to that followed in the conventional practice; (ii) this technique was economically and technologically feasible; (iii) no negative effects on the yield and or quality; (iv) gradual reduction in nitrate leaching into groundwater, from nitrogen fertilizer applied to the crop, as evident from gradual reduction in nitrate concentrations in the groundwater.

Therefore, the BMP developed met all the criteria specified in the Nitrate BMP bill. Furthermore, progress and proactive attitude of the farmers and the leadership of FL DACS and the BMP research team contributed to high participation by all parties at all phases of the BMP development and rule making process which ultimately translated into high level of voluntary participation by the producers in acceptance of the BMP. This level of success would have been virtually impossible if the BMP was implemented by the regulatory agency without the active participation of all parties including the end users. The proof of adaptation of the BMP was to be documented by the producers with respect to all requirements of the BMP.

This nutrient management legislation was further expanded in 2003 to include phosphorus BMP due to evidence of phosphorus contamination of surface water. This legislation enabled collection of additional tax from fertilizer dealers. The funds collected were used for research, development, demonstration, and implementation of suitable interim measures, best management practices or other measures used to achieve state water quality standards for nitrogen and phosphorus. These funds were also used for cost sharing grants, technical assistance, implementation and tracking of BMPs, and conservation leases or other agreements for water quality improvements. This approach used by the FL legislature is a fine example of revenue generation from the source of pollution to develop and implement strategies and or practical solutions to mitigate future pollution.

FL DACS was required to consult with FL DEP, Department of Health (DOH), water management districts (WMD), environmental groups, the fertilizer industry and representatives from the affected farming groups throughout the process of developing, approving, and adapting fertilizer interim measures, BMPs or other measures. This is an important factor that contributed to the success of the program. Monitoring and verifying the effectiveness of adaptation of the interim measures or best management practices was carried out by the FL DEP, with cost of such activities being reimbursed by FL DACS using the same pool of funds collected from fertilizer dealers. The fees collected for the sole purpose of developing and implementing N and P BMPs under this program includes: (i) $100 for each license to distribute fertilizer; (ii) $100 for each specialty fertilizer registration; (iii) $0.50 per ton of all fertilizers containing N or P sold in FL.

17

United States of America - Farm Bill

The first Farm Bill was passed in 1933, during the great depression, called as Agricultural Adjustment Act (AAA). The intent of the bill was to provide financial assistance to farmers to keep the land fallow as price control to avoid glut in the market which lowers the food price thus affects the net returns to farmers. Price control is necessary so that farmers can continue farming to assure adequate availability of food products. Farm Bill was necessary to provide adequate funding for the government to buy excess food from farmers at a fair price without hurting the farmers return and government can sell the food at a later date if the overall food supply in the market was reduced by bad weather or other factors impacting food production. Therefore, the overall intent of Farm Bill was in the best interest of the general public to assure food supply while ensuring farming as an economically profitable enterprise. This bill also included a nutrition program, which later became known as 'Food Stamp'. The subsequent Farm Bill of 1938 established a permanent program with a requirement to renew Farm Bill every five years.

Major changes in Farm Bill were introduced in 1996 with intent to eliminate subsidizing farmlands and purchase of excess grains from farmers. This was designed to allow the free market to dictate the food price. Instead, government introduced crop insurance program to receive farm payments. In late 1990's Farm Bill introduced direct payments to struggling grain farmers as an incentive to continue grain production, i.e. staple food source. This resulted in grain farmers to receive annual payments based on the yield and farm area recorded in the previous decade.

Farm Bill of 2008 was called 'Food, Conservation, and Energy Act', with an annual budget of $100 billion with nearly 80% allocated to food stamp and other nutritional programs. The 2008 Farm Bill annual spending increased to $288 billion, which raised serious concerns with respect to increasing the budget deficit. The Bill included subsidies to biofuel crops, the program which was thought to contribute, in part, to a significant increase in food prices. The 2008 Farm Bill budget was 47 percent greater than that of 2003 Farm Bill. It was highly controversial due to the fact that in the past 10 years 75% of the total farm subsidy went to only 10 percent of the farmers, therefore, was criticized for helping a small portion of the farmers.

During preparation of the 2012 Farm Bill there were considerable concerns on the increasing size of the budget, about $768 billion over 10 years, with major portion of the budget for nutritional program and direct payment to farmers. This combined with overall budget deficit and public and media calls for reductions in Federal spending elevated the negative sentiments on the nutritional program and farm subsidies. The Bill failed to pass in the House, therefore, to keep the programs active 2008 Farm Bill was continued one more year under the 'American Taxpayers Relief Act' of 2012.

The normal cycle of Farm Bill has is five years. About 2 to 3 years prior to expiration of the farm bill discussions on the new farm bill begins. A number of town hall meetings are organized to discuss the critical issues in preparation for a bill that comes up for voting in both houses. Serious disagreements in the bill between the two houses are resolved by congressional conference committee. The bill has to be passed again by both houses and become a law when signed by the president. When the farm bill was originally created in 1933 as a part of the Agricultural Adjustment Act, control of market prices of major agricultural commodities was the main emphasis. Appropriated federal funds were used (under Farm bill) to pay farmers not to produce some main crops in order to maintain a stable market price. These commodities were: corn, soybean, wheat, rice, and peanuts. The subsidy for market price regulation was also extended to other commodities such as dairy and tobacco. The underlying philosophy was to maintain certain level of production to minimize market driven price fluctuations. The commodity price drops with surplus production. In order to ensure a given target production level of a given commodity, some farmers had to be paid not to produce that commodity, i.e. those farmers leave the farm fallow but get the compensation from the government under this program for the loss of farm income due to fallow.

The 2008 Farm bill was $300 billion in mandatory spending; of which 62% was for nutrition program (supplemental nutrition assistance program, SNAP), 14% for crop subsidies, 9% conservation programs, and 8% for crop insurance.

Farm Bill Distributions: 2008- 13

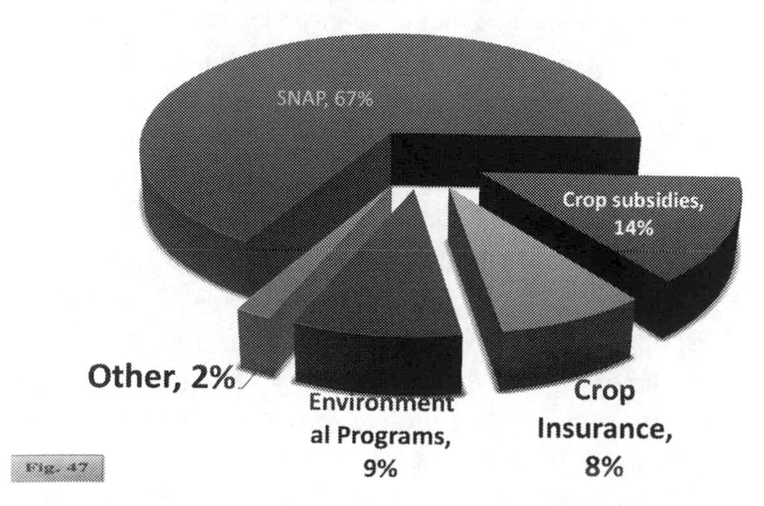

Distribution of Farm bill (2008-2013) funds across different programs. Notice 67 percent of the funds go to supplemental nutrition assistance program (SNAP), formerly referred to as food stamp.

The remaining 2% funded farm credit, rural development, forestry, energy, livestock, and horticulture/organic agriculture. Therefore, a very large portion of farm bill funds were used for supplemental nutrition program generally known as food stamps, i.e. nutrition assistance for low income people. The main driving force for SNAP is to ensure providing basic food and nutrition needs of the children when the families are experiencing economic difficulties. Therefore, it is a misconception that farm bill is government hand out to farmers. The SNAP benefits in 2010 increased to $ 64.7 billion as compared to $34.6 billion in 2008. Almost doubling of this benefit in just two years has been extensively debated, due to rapid increase in enrollment, in part due to increasing economic hardships faced by many more households. This is an indication of the bad economy and job market in the last few years which forced many households to seek government assistance for basic needs. Often the farm bill is criticized by the mass media as an unnecessary government payment to rich farmers. This criticism is unjustifiable considering nearly 2/3 of the farm bill funds go to SNAP. The conservation program and crop insurance are needed to maintain environmental sustainability of the natural resources and to help farmers to ride out the natural disasters. Only 14% of the total farm bill funding goes to crop subsidies to maintain stability in commodity prices so that farmers can expect reasonable net returns by avoiding glut in the market that can lower the commodity price and net returns. With increasing emphasis on reducing Federal Budget during the recent years, the crop subsidies, crop insurance programs are closely monitored for potential reductions. These two programs combined accounts only for 1/5 of the total farm bill budget. The lack of long term budget management strategy and partisan politics are not conducive to make any changes in mandatory programs.

18

US Agriculture Statistics

Annual gross output from US agriculture is over $370 billion and supports about 750,000 employments. Demands for export of American agricultural commodities provide a strong economic base. Despite shrinking agricultural sector in the US, the technological advances and increased production efficiency makes the US agriculture envoy of the world. As a result despite < 2 percent of the US population engaged in agricultural enterprise they manage a highly efficient production system to feed the entire US population and also produce more to support a vibrant export market. In 2012, the value of agricultural export attained $136 billion, and projected to increase in the future. About 36% of the gross farm income can be traced to export income. Rising disposable income of middle income households in the emerging economy countries in the world have contributed to increasing demand for US agricultural commodities and products.

The per capita food expenditure in the US increased from $3600 in 1985 to $ 4230 in 2011. The per capita income increased at a much faster rate in this period. As a result the percent of household income used for food decreased from 15% in 1984 to 13% in 2009. With increasing household income the consumers demand greater quality food, i.e. better quality of meat, a variety of fruits and vegetables, organic and specialty food items.

There has been an increasing demand for organic food. The retail value of the organic industry grew to $32 billion in 2011, as compared to $21 and $4 billion in 2008, and 1997 respectively. Acreage under organic production grew by an average of 16.5% during the 2002 to 2008. However, the sale of organic produce in the US is still only about 3% of the total sale of all food. Increasing emphasis on using 'locally grown' products movement has been a factor in increasing organic production. Organic farming in most cases is small scale and often supplies local market, hence readily being classified as 'locally grown'. Roughly 5% of the US farms acreage is classified as 'locally grown'. This is usually small farms, i.e. < $50,000 gross annual sales.

19

Agricultural Research and Development (R&D)

Continued investment and growth in agricultural R&D are critical to boost productivity and net returns. Some recent estimates reveal for every dollar invested in agricultural R&D efforts produce 10 to 20-fold greater returns in the form of benefits to the society. The public and private investment for agricultural R&D increased from $6 billion in 1970 to about $11 billion in 2007, i.e. 83% increase in 37 years which is roughly 2% increase annually. On an annual basis, the R&D investment is roughly 50% each from public and private sources. The rate of increase was also very similar in both public and private sources of R&D investments.

20

History of USDA Role in Research

Agricultural research in the US is a joint venture between the public and private sector. Major players of public sector support are Land Grant Universities (LGU) in every state, and agricultural research service (ARS) agency of the US Department of Agriculture (USDA). The funding for LGU is primarily from the state government plus some formula funds from the federal government. The role of USDA in support of agriculture research is twofold: (i) support intramural research through ARS, and (ii) execution of competitive grant program which mostly supports scientists from the LGU, through another agency called National Institute for Food and Agriculture (NIFA). The annual appropriated budget for each agency is approximately $1billion. Therefore, total Federal investment for agricultural research is about $2 billion per year.

The philosophy of ARS is to fund stakeholder's driven, long term, high risk research. This kind of research is not amenable to competitive grant program or private sector support; therefore, it was necessary to make adequate and continued Federal funding available to support such research. Although this was a novel goal and maintained for a long time up until few years ago when federal government budget came under increasing scrutiny which resulted in flat or decreased funding during the last several years for ARS. To cope with the current budget strain, the ARS began to encourage scientists to seek for external grants to support their mission driven, stakeholder priority based, long term research. This is sometimes difficult due to the fact that the problem solving research is often applied research projects and are long term hence do not meet the evaluation criteria set for the competitive grants. After several years of level funding, the FY 2013 budget reduction was about 8% due to across the board budget reductions, i.e., sequestration. Net outcome of all recent budgetary strains have seriously impacted many long term research programs carried out by the ARS.

National Institute for Food and Agriculture (NIFA) on the other hand is primarily granting agency. The funding is earmarked to various priority research areas, and request for proposal (RFP) goes out various times during the year to solicit research proposals. The proposals are reviewed by an expert review panel and the top ranking proposals within each priority areas are funded, subject to available funds for each category of research. More recently, NIFA has developed a strategy of funding large grants which involved multi-disciplinary, multi-state, multi-agency cooperation. Due to the level of funding for such large proposals the number of projects which are being funded decreased considerably. The basic philosophy of this approach is to encourage cooperation across different disciplines, from multiple states on a regional basis. Such cooperation is looked upon by the granting bodies as innovative and can bring about major impacts on funding investment. The critics of this philosophy indicate that such major grants don't guarantee effective, needed, major impact deliverables. Regardless, all indications are to promote increased cooperation among scientists from different disciplines and emphasize national or regional scale projects, rather than investment of research funds to address local problems by a small group of scientists without multi-disciplinary cooperation. Most of the NIFA funding goes to LGU scientists. ARS scientists can only serve as cooperators, as of now, except for some special grants. The LGU mission covers research, teaching, and extension. The current trend often is to encourage inclusion of extension in most research proposals. This ensures emphasis on transfer of technologies to end users on a timely basis and promotes increased involvement by the producers and stakeholders in all stages of project planning and execution.

USDA was created in 1862. The LGU system was authorized by Morrill Act to facilitate teaching of agriculture in each state. In 1868 USDA began research on animal diseases. Foot and mouth disease was first reported in the US in 1870. In 1873, USDA secured Washington Navel oranges from Brazil and was first introduced to California. The introduction of Wheat from Turkey into Kansas occurred in 1874. The first Agricultural Experiment Station was established in 1875 in Wesleyan University Middletown, CT. First milking machine was invented in 1878. In 1887, Hatch Act was passed which provided Federal funds to State Agricultural Experiment stations.

The Department of Agriculture gained the cabinet status in 1889. In 1890, the second Morrill Act was passed to authorize land grant colleges for African Americans. These colleges of which some attained the University Status later are commonly referred to as 1890 college/university or Historically Serving Black College and University (HSBCU). In 1894, 'Carey Land Grant Act' granted land to the western states after the states provided irrigation. Preservation of germplasm of various crops has been and continues to be a major mission of the USDA. This program supports collection of germplasm from around the world; multiply the seeds, store, and distribute the seeds to researchers. The germplasm bank is safe storage of the seeds in multiple locations for future use. This program originated in 1898. Currently germplasm of the most crops are stored in safe storage.

Soil characterization or classification is also another major mission of USDA. This is commonly called as soil survey or mapping. The soil survey reports are in the past printed for each county. More recently these are digitized by the Natural Resources Conservation Service (NRCS) agency of the USDA. This provides a very useful guide for anyone to get a basic knowledge of the soil with reference to its suitability for various uses including agriculture, housing, road construction or other uses. This process began in 1899. Meat inspection act was passed in 1906.

Farm Bureau is an organization for promotion of agricultural interests. This organization provides support to farmers and ranchers and also is a lobbying force to promote the interest of agriculture. The first office of the Farm Bureau was opened in 1911, in Boom County, NY.

Agriculture extension is a service to farmers and ranchers on various production and land management issues. This is a bridge between the researchers and the end users of the research recommendations. The role of extension is to disseminate the research deliverables to the farmers in a practical and non-technical language. The philosophy of extension is to simplify the research deliverables in a user friendly form that can be followed by the farmers so that the recommendations will be readily adapted. This is generally accomplished by a publication of 'fact sheets', 'extension bulletins' or by direct presentation of the research results in farmers meetings. The origin of extension program dates back to 1914 with passage of 'Smith Lever Act'. This act authorized the third mission of the Land Grant system, i.e. extension. During the recent years however, extension wing of the LGU system is often closely scrutinized, downsized, and also eliminated in some states. The non- agricultural population is questioning the validity of using public funds to provide this free service to farmers and ranchers. As the agricultural sector population continuously dwindles due to Urbanization, several states legislature base is predominantly urban and gradually loosing support to agricultural extension service.

In 1917, USDA issued U.S. grade standards for potato. This was the first official grade standard for fruit and vegetable and continues to be the industry standard for evaluation of tuber size and quality. Grading of fruits and vegetables based on the size and deformity was required to determine the premium price for the product.

The control or eradication of disease causing organisms is generally done by spraying chemicals, i.e. pesticides or fungicides.

This is an effective and required control measure needed to minimize crop losses from diseases and pests in order to maintain high production and product quality. However, since the 1960's there has been an increased interest by the consumers on the impact of the chemical residue on food safety as well as on the environmental quality. Some chemicals began to leach in the soil below the root zone and ended up in the groundwater. Some chemicals contaminated the surface water bodies when mixed with surface water runoff. These negative impacts of routine use of chemical and nutrients, commonly referred to as 'non-point source pollution', resulted in somewhat negative perception on the harmful effects of routine agricultural practices. Proper management of chemical application, with respect to timing and intensity, as well as timing of irrigation has an effect on the leaching losses of chemicals below the root zone. This in turn can result in chemical contamination of ground water. Hence the origin of non-chemical techniques to control of crops diseases and pests. A pest can be controlled by another insect that can feed on the pest, i.e. predators. A pest can also be controlled by or use of virus, bacteria or fungus that can cause disease only to the pest but no harm to the host plant, humans or the environment. This process of controlling the disease and pest is commonly referred to as 'Biological Control' or short form 'biocontrol', which began in 1919, and is considered as sustainable practice by way of reduced use of chemicals which, in turn, mitigates the residue of chemicals in the produce and minimizes contamination of chemicals to surface or groundwater.

Inspection of live poultry by USDA began in 1926. Tillage of the soil is a prerequisite for crop production. Tillage is done before planting a crop to create an ideal environment for the seeds to germinate and for the crops to grow to produce high yields. After the product of economic value is harvested the remaining portion of the plants, i.e. biomass, has to be incorporated to the soil so that the biomass decomposes and the nutrients in the biomass is released and made available to the subsequent crop. This process is referred to as 'mineralization' of nutrients from stable organic form into readily available forms. The soil disturbance caused by tillage resulted in soil erosion; in some sensitive areas subject to soil, landscape, and climatic conditions. The fine soil particles are blown in wind resulting in air pollution. In high rainfall areas the soil particles are carried by excess water runoff from grower field that is carried to canals, and stream and eventually to the rivers. This is a major environmental problem and also degradation of the soil since the topsoil is transported, the loss of topsoil contributes to reduced soil productivity. The first official recognition of major impacts of soil erosion on soil productivity was in 1928. This became the impetus for establishing the Soil Erosion Experiment Station. In 1933, 'Soil Erosion Service' (SES) was established which later became known as 'Soil Conservation Service' (SCS) and subsequently renamed as 'Natural Resource Conservation Service' (NRCS).

The US was hit by the worst drought in 1934, which impacted >75% of the country.

Effective insecticide is the key to control insects to minimize crop damage. Some insecticides are potent and kill the insects on contact. The other groups of insecticides are called 'systemic', i.e. the mode of operation is not by contact, instead when applied to the soil it is taken up by the plants, yet not harmful to the plants, carried to the foliage via xylem transport. When the pest feeds on the foliage the chemical kills the pests. The origin of this concept of systematic insecticide was in 1936, which was demonstrated by application of selenium to the soil to control aphid. Selenium was absorbed by the roots, transported to the foliage which killed aphids. This is an efficient way to control target pests without affecting the beneficial insects such as predators. Contact insecticide on the other hand, kills all insects which come in contact.

Historically most plants are grown in soil. Function of soil is to provide support, an environment to retain water which was subsequently made available to plant roots to absorb, and also media containing plant nutrients either from plant residue or by application of various nutrient sources such as fertilizer and other amendments.

Therefore, soil was considered as an appropriate medium for plant growth. This concept gradually changed by introduction of 'hydroponic' system in 1929, in this system of plant growth the plant top is supported by some support mechanism while the roots freely suspended in a solution containing all plant nutrients. The solution was aerated by compressed air. The solution is carefully monitored for depletion of nutrients and the amount depleted was replenished to provide constant supply of all nutrients at optimal levels needed for plant growth and production. The aerated nutrient solution became the source of all nutrients and air, therefore practically serves as a growth medium as far as plant is supported properly. This in some way was very efficient for high production, because the nutrient availability can be carefully monitored and regulated unlike the soil medium. Hydroponic system is very convenient for greenhouse grown plants and the growth media can be altered to support desirable production targets.

Another soilless culture is by using vermiculite which originated in 1940. Unlike the hydroponic in Vermiculite culture, this provides plant support. The nutrients are applied by nutrient solution as needed. The vermiculite medium provides adequate aeration for root growth so no need for any artificial air circulation. This is also a preferred method of growing plant in greenhouse in pots.

Weed control is a much needed operation in production of any crop plant of economic value. Competition by the weeds hinders growth and production of crop plants. Historically when labor cost was low hand weeding was preferred route because of small land area, and equipment and chemicals for specific weed control were not available. This is almost impossible in modern agriculture where large holdings are the norm, labor is expensive and minimizing weed competition is required to maximize net returns. The concept of using a chemical that can selectively control the growth of weeds without affecting the crop plants began in 1942, by the discovery of mode of action of a common herbicide (2, 4-Dichlorophenoxyacetic acid, abbreviated to 2,4-D) which became the most important herbicide subsequently.

In 1945, United Nations established Food and Agriculture Organization (FAO).

Agricultural Research Service (ARS), an intramural research agency of the USDA was established in 1953. This provided Federal funds for agricultural research in about 100 labs across the country to conduct long term, high impact, mission driven projects often in collaboration with Land Grant Universities – research and extension mission.

Improved varieties of alfalfa, lettuce, peach, bermudagrass, and potato were released in 1953. Most notable among the release were 'Vernal' alfalfa, important for the arid region, and 'Red Lasoda' potato, most popular red skin variety.

In 1954, a landmark handbook, entitled 'Diagnosis and Improvement of Saline and Alkali Soils', was published by the Agricultural Research Service agency of the USDA. This handbook was intended as a guide for identification, properties, and management of these problem soils. These soils pose limitations for crop production due to high concentrations of soluble salts (saline soils), or exchangeable sodium (alkali soils), or both (saline – alkali soils). This Handbook is the most comprehensive guide used around the world for management of these problem soils.

Safe, long term storage of plant germplasm for future use is a major mission of the USDA-ARS. Scientists from ARS travelled to various countries for selection of germplasm for introduction into US for commercialization depending on the need and adaptability. Likewise, germplasm of most crops are stored in the germplasm storage centers. Scientists from around the world can request a small sample of germplsam for seed multiplication for their research for any trait of specific interest. The first National Seed Storage laboratory was established in 1958.

Monitoring soil water content is needed to evaluate the status of soil water content available to plants so that water deficit can be adjusted by irrigation in irrigated production system. This can be done by taking the soil sample and estimation of water content in the soil by weighing method (gravimetric method) before and after oven drying at 100 degree C for 24 to 48h. This method is time and labor consuming. Therefore, the need for instant in-situ measurement of water content. In 1961, neutron probe was developed for instant measurement of soil water content at various depths. This is a spot measurement which precludes the evaluation of changes in soil water content on real time basis. Significant progress was made subsequently to enable real time measurement of soil water content at various depths and automation of this measurement and link this technique to trigger irrigation which can be controlled remotely to reduce the labor and avoid any mild water stress effects on the plant growth and production.

In 1962, Rachel Carsen's book 'Silent Spring' was published documenting the negative environmental impacts of pesticide use in agriculture. The origin for this book was the incidence of bird kill supposedly linked to use of DDT for mosquito control. Her book instigated wide debate on the environmental consequences of pesticide use, and played a role in nationwide ban on the use of DDT.

An important parameter for managing plant water relations is loss of water from the plant, termed 'transpiration', and that from the soil, i.e. evaporation. These two losses combined is called 'evapotranspiration (ET)'. Irrigation management is aimed to replenish the ET deficit either full or to a tolerable level without resulting in major yield losses. There are some techniques to estimate the crop ET. Direct measurement of ET is by weighing technique, wherein the plant is grown in a tank with soil to the depth of root zone, with appropriate attachments to collect the water leached below the root zone. Given the known amount of water applied and with the measurement of water leached below the root zone, the weight loss of soil plus the plant indicates the loss of water by ET. This requires a specialized precision instrument termed weighing lysimeter. To make this measurement on a real time basis, this instrument should have automated weighing and data processing capabilities so that ET can be estimated over the entire growing period. The first such precision weighing lysimeter was developed in 1962.

In 1970, Environmental Protection Agency (EPA) was established to address and mitigate both industrial and other sources, including agricultural, of pollutants. In 1972, clean air act, clean water act, and consumer product safety acts were passed. Safe drinking water act was passed in 1974. This act authorized EPA to establish minimum standards for drinking water quality to be complied by all public drinking water authorities. This law applies to all water sources, underground or above ground, used or designated as potential future use for drinking. The 1996 amendment to this law required EPA to consider risk and cost assessment, and best available peer reviewed science while developing these drinking water quality standards.

Evaluation of livestock feed quality is important to ensure that high quality of feed is maintained in the forage market. The detailed traditional forage quality analyses are very tedious and time consuming. In 1976, rapid, non-destructive technique for forage quality analyses was developed by using near-infrared reflectance (NIR) spectroscopy.

All through the 1980s and 1990s there have been a lot of debates on the role of GMO (genetically modified organisms) in crop improvements. GMO is a plant developed by placing a copy of desirable gene or genetic material from one plant or an organism into another plant. This process is used to enhance the tolerance/resistance to diseases and pests, or enhance crop quality or nutritional value etc. The plants developed by this process are also called 'transgenic' plants. This term was first used in 1980 for mice that carried a new and recently introduced gene. Currently GMOs are commercially available in the US for corn, soybean, alfalfa, papaya, summer squash, canola, cotton, and sugar beet. Much of the debate about pros and cons of GMO are often emotional and lack of full understanding of the issue rather than based on credible scientific information. To some extent the scientific breakthrough failed to bring the general public along for sharing the information. This in turn created mistrust and was the origin of emotional uproar and sentiments against GMOs. This is an extensive highly emotional topic not within the scope of this book. The brief description of this topic in this section was only to highlight its scientific breakthrough and capabilities.

The first genetically modified crop plant that was released was tomato in 1982. In this case, a gene responsible for producing certain enzyme (called polygalacturonase) that was needed for fruit softening was deactivated. The rationale for this was to allow tomato long shelf life, therefore, develop full flavor, hence called 'FlavrSavr'. Despite the GMO tomato being approved in US, by 1998 these tomatoes disappeared rather fast from market, in part, due to public concerns on the perceived damage to environment and ecosystem by the GMOs. Currently, no GMO tomatoes are commercially available in US or Europe. This is an example of public mistrust in application of advanced technology for food system, and failure to share research based information with the general public by the scientists and multinational companies involved in the development of GMOs.

Biological control (also called biocontrol) is a collective term used for non-chemical control of pests and diseases. This is done by beneficial insects or organisms feeding or infecting harmful (to crops) pests or disease causing organisms thereby control the damages. The former however, pose no threat to the crop in question in a given production system. In an effort to reduce the chemicals (used for control of diseases, pests, and weeds) foot print on the crop products, there is a lot of interest in practical application of biocontrol technique. A significant progress in this effort occurred in 1981 with development of techniques for first commercial application of nematodes for control of carpenter worm in commercial Fig orchards in California.

Conservation reserve program (CRP) is administered by Natural Resources Conservation Service (NRCS) Agency of the USDA to protect excessive soil erosion of highly erodible land. This is done by designating some land for year round grass and tree production to avoid cultivation which can promote increased soil erosion. In some productive land which needs to be cultivated for production of annual crops, soil erosion can be minimized by conservation buffers. This can be in the form of grassed waterways, or contour grass strips that can reduce or eliminate sediments and nutrients runoff into water bodies. This program promotes removal of environmentally sensitive land from production by providing permanent grass or tree vegetation to improve soil health, reduce sediment and nutrient contamination of water bodies, and support wildlife. The farmers who are required to take up these conservation measures will be compensated by payment as incentive. This program began in 1985 and continues currently under the Farm Bill.

In 1986, Germplasm Resources Information Network (GRIN) was established. This program is administered by the USDA – Agricultural Research Service agency, and is the world's most comprehensive database of agriculturally important plants, animals, microbes, and invertebrates. The National Genetic Resources Program (NGRP) was approved by the US Congress in 1990, which authorized the NGRP to acquire, characterize, preserve, document, and distribute to worldwide source of germplasm resources for food and agricultural production.

Soil moisture and temperature are most important characteristics for optimal management of agricultural soils. Traditional methods of measurements of these parameters are very cumbersome, time consuming, and labor intensive for real time measurements. In 1993, significant advance was made when remote sensing technique was applied for measurement of soil moisture and temperature.

Simulation model is a technique to predict the crop response as well as fate and transport of agricultural chemicals and nutrients applied for routine crop production. This is certainly economical as compared to measurement of responses by conducting such studies in each soil type and crop production conditions. These models are developed by using the response data from variety of situations representing the large variations. Therefore, given the basic soil properties and climatic conditions, we can use the model to predict the crop response and changes in soil and water properties. This is a good planning tool so that appropriate changes in management practices can be incorporated to maximize efficiency of input use, and minimize negative environmental impacts, such as deterioration of water and/or air quality. With increased attention to sustainable crop production, there has been greater need to predict the fate and transport of agricultural chemicals and nutrients in a given production condition, so that appropriate management options (best management practices, BMP) can be utilized to minimize negative environmental impacts. Among various models of this utility, Root Zone Water Quality Model (RZWQM) was developed by the USDA – ARS scientists and made available for the public use in 1995. Since then this model has been used by researchers, consultants, and policy makers in understanding the impact of current production practices under variety of production conditions and developing management options to mitigate negative environmental impacts.

Remediation of contaminants in soils, sediments, surface water, and groundwater is a necessary task for restoration of these natural resources. If remediation is not done and the land is used for agricultural production, there is potential for contaminants to enter into human food chain leading to significant damages. For several contaminants, there are no easy techniques to remediate large areas and enable use of the land for crop production. A new technique to use of green plants to stabilize or reduce contaminants in soil is termed as 'phytoremediation'. Once the contaminants are taken up by these plants, they have to be harvested and disposed off from the site of contamination. Based on in depth research various plants have been identified, by 1999, as effective for removal of specific contaminants. The success of this program is dependent on: the ability of the plants to degrade or uptake contaminants, adaptations to local climates, high biomass production, distribution of roots, ease of planting and maintenance, and adaptability of the plant to given climate in the contaminated site etc.

21

Changes in China

With increasing globalization the economy of each country is interlinked to a great extent. Fluctuations in economy in remote corner of the world have immediate and long lasting ripple effects on the rest of the world. As second ranking economy of the world, China (GDP of close to 11 trillion dollars, is approximately 58% of that of the No. 1 ranking country, i.e. USA) is a major player in impacting the world economy. Unlike the western countries, agriculture contributes about 11 percent of the gross domestic product (GDP) of China.

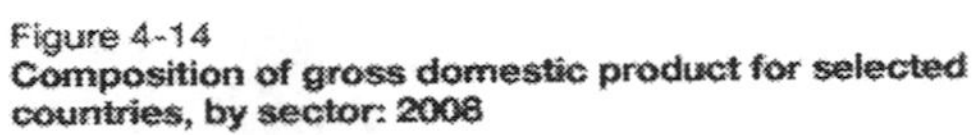

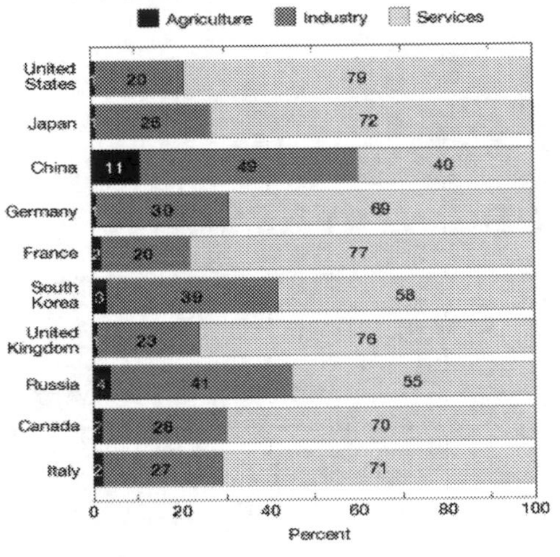

NOTES: Data for Russia are 2007. Data cover the 10 largest R&D performing countries.

SOURCE: Central Intelligence Agency, *The World Factbook*, http://www.cia.gov/library/publications/the-world-factbook/index.html, accessed 25 February 2009.

Science and Engineering Indicators 2010

Fig. 48

Percent contribution of industry, agriculture, and services sector to total gross domestic product (GDP) of 10 countries.

In contrast in many of the western countries agriculture contribution to the total GDP is <2 percent. The corresponding percentage for Russia is about 4 percent. Services represent a large portion of the total GDP (70 to 80 percent), while industrial output constitute 20 to 30 percent for most western economies and Japan. For China, Russia, and South Korea, services and industry contribution of the country total GDP are in the range of 40-58 and 39-49 percent, respectively.

For a very long time, China tried to maintain self-sufficiency in food production in the interest of preserving food security. However, the high economic growth of the country contributed to increased discretionary income which resulted in gradual shift in food system and increased imports of grains and land intensive crops. Since 2001, agricultural trade between China and US increased rapidly and by 2012-2013 the import of agricultural products from the US attained almost 26 billion dollars,

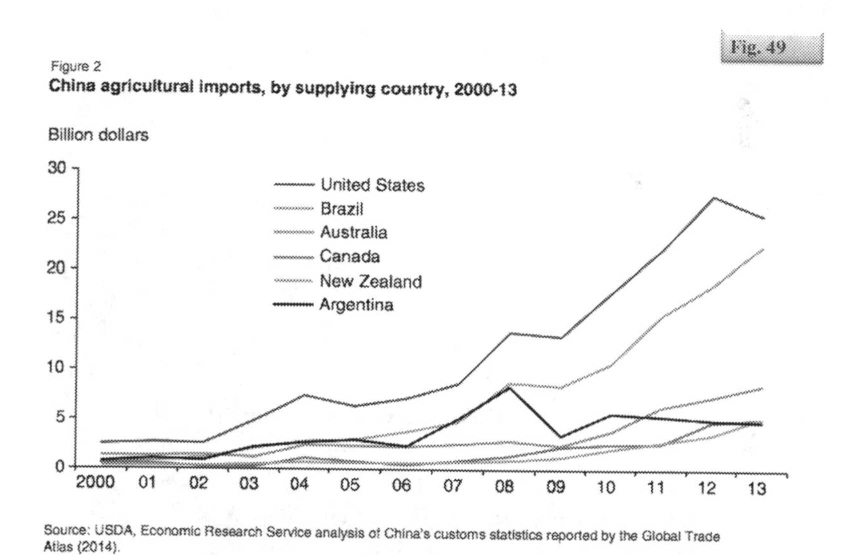

Trends in import of agricultural products by China from different countries during 2000 through 2013

which represented the largest share of import of agricultural products by China, followed by that from Brazil. Import of agricultural products from these two countries combined accounts $ 48 billion. The rest of the import originated from Australia, Canada, New Zealand, and Argentina each accounted for $ 5 to 7 billion. By 2012 and 2013, the US agricultural export to China reached $26 billion, which was two – fold greater than that in 2008.

Fig. 50

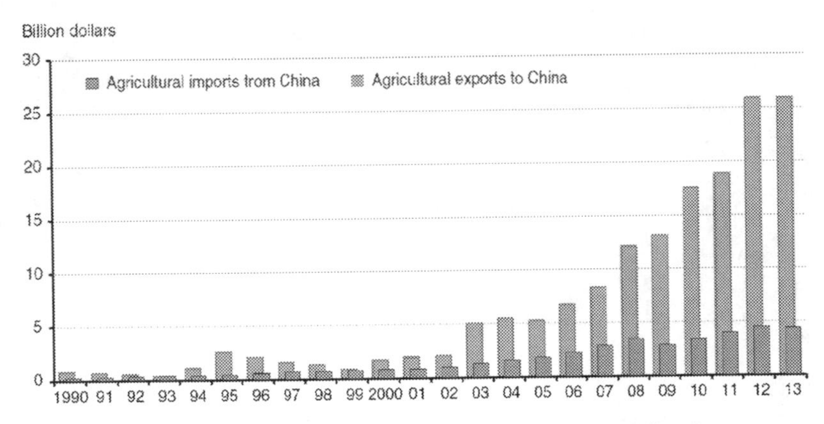

Reproduced with Permission: USDA – Economic Research Service

Trends in United States of America agricultural import
from China and export to China from 1990 to 2013.
Notice the steep increase in export post 2003.

During 2000 – 2013, the agricultural export from US to China increased quite rapidly as compared to the agricultural import from China to US. By 2012, the US export of agricultural products valued almost 6-fold greater than that of agricultural import from China to US. This once again supports the importance of China as a major US partner in agricultural trade, where US has maintained distinct trade surplus with China, unlike in many other sectors where US trade with China shows deficit.

In 2008, the agricultural export from US to China was about $12 billion, which ranked fourth behind Canada, Mexico, and Japan. By 2013, the US export of agricultural products to China doubled, and China became the major US trading partner for agricultural products ($ 25.9 billion), far ahead of other trading countries, such as Canada ($ 21.3 billion), Mexico ($18.1

billion), Japan ($12.1 billion), and European Union ($11.9 billion) (Table 6).

Top destinations for U.S. agricultural exports, 2008 and 2013

Country/region	2008	Country/region	2013
	bil$		*bil$*
Canada	16.3	China	25.9
Mexico	15.5	Canada	21.3
Japan	13.2	Mexico	18.1
China	12.1	Japan	12.1
European Union-28	10.1	European Union-28	11.9

Source: USDA, Economic Research Service analysis of data from USDA, Foreign Agricultural Service, Global Agricultural Trade System.

Reproduced with Permission from: USDA – Economic Research Service

Table 6.

The main agricultural products imported by China are oil and oilseeds (40%), cotton (13.4%), and grains and feeds (12%).

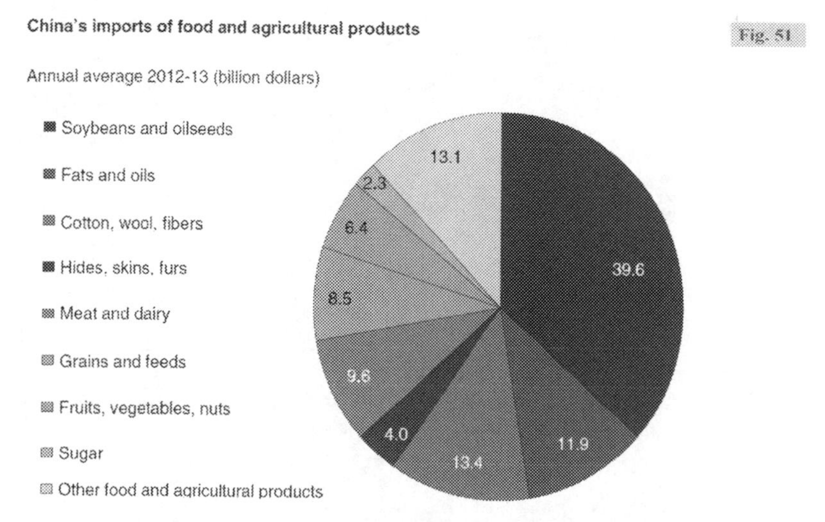

Import of different food and agricultural products by China (in billion dollars) for 2012-2013.

US is a major exporter of animal feed to China, such as distiller grain, a byproduct of ethanol production from corn grain (Table 7).

With China's economic boom and changing food habits, China is becoming a growing market for high valve agricultural products such as alcoholic beverages, dairy products, bread, extracts of coffee and tea (Table 7). With the increased production of corn based ethanol in the US, there became an abundant production of byproduct dry distillers grain. China emerged as a major importer of this byproduct to feed their growing animal industry. The net value of import of this byproduct by 2013 was about $1.1 billion.

U.S. share of China's leading agricultural imports, 2012-13

Item	Average Chinese import value	U.S. share of China's imports	U.S. rank
	$ Billion	*Percent*	*Number*
All agricultural products	109.0	24	1
Soybeans and other oilseeds	40.6	36	1
Fats and oils	11.9	2	11
Cotton	10.1	30	1
Meat	5.0	25	1
Cereal grains	4.9	42	1
Dairy	4.2	10	2
Fruit and nuts	3.9	13	4
Wine and beverages	3.1	3	7
Cattle hides	2.6	53	1
Wool	2.7	<1	13
Baking products	2.3	4	13
Vegetables	2.5	2	5
Sugar	2.5	5	5
Fish meal	1.7	15	2
Distillers' dried grains	1.1	99	1
Tobacco	1.4	11	3
Live animals	.5	15	3
Hay and forage products	.2	95	1

Reproduced with Permission from: USDA – Economic Research Service

Table 7.

Following 2005, there has been a rapid increase in import of most agricultural products, particularly for cheese, cookies, bread and pastries products.

Fig. 52

China's imports of selected processed food products, 2000-13

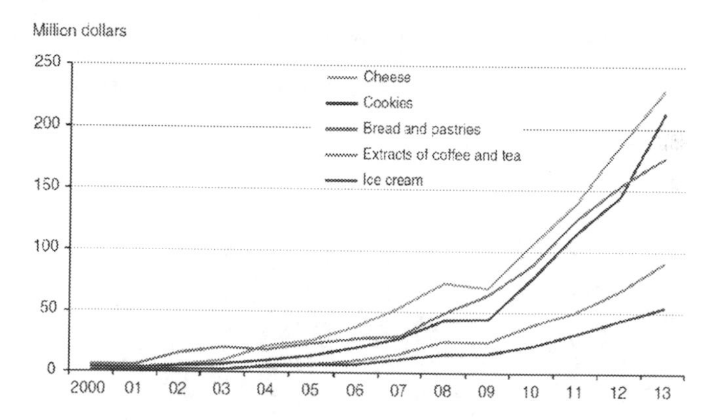

Reproduced with Permission: USDA – Economic Research Service

Trends in import of some food products by China during 2000 – 2013.

Similarly the import of milk products, and meat also increased since early 2000 to current years.

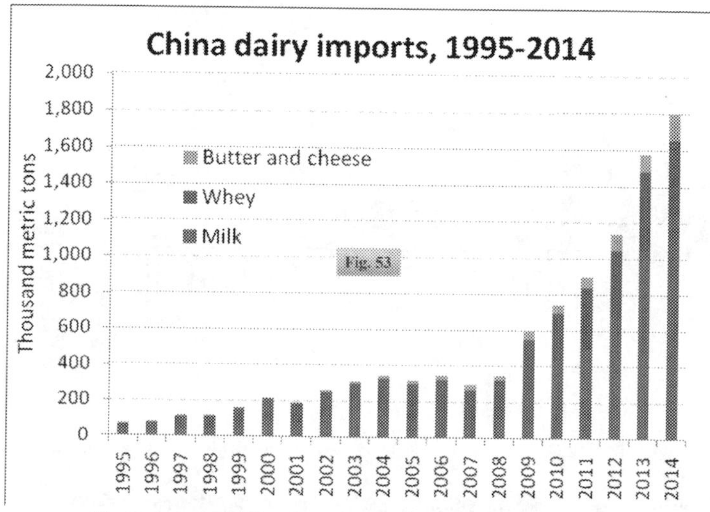

Reproduced with Permission: USDA – Economic Research Service

Trends in import of dairy products by China during 1995 to 2014.

Fig. 54

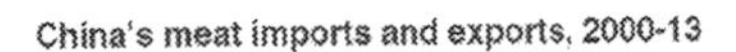

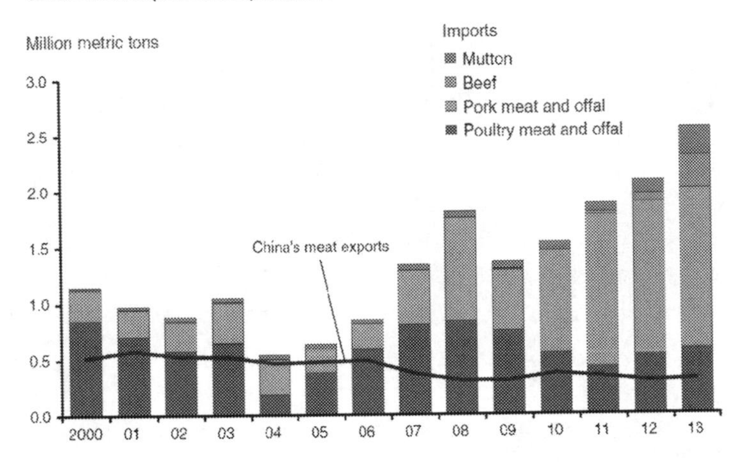

Reproduced with Permission: USDA – Economic Research Service

Trends in import of meat products by China during 2000 to 2013.

China's net trade in grains, 2000-14

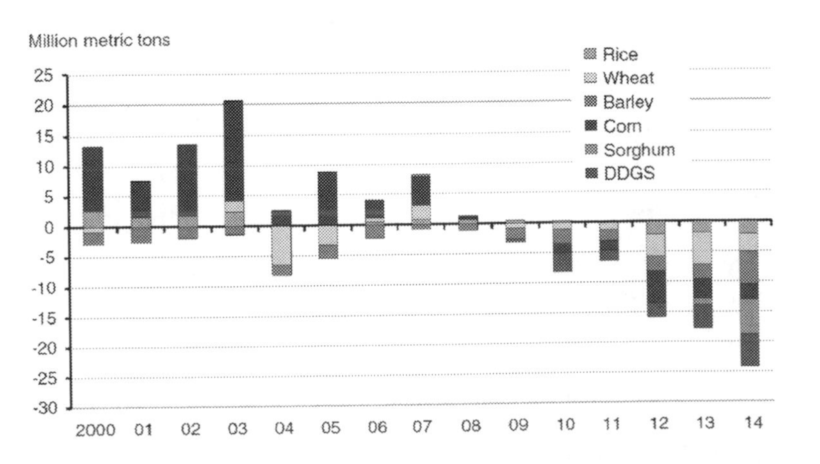

Reproduced with Permission: USDA – Economic Research Service

Trends in China's net grain trade during 2000 through 2014.

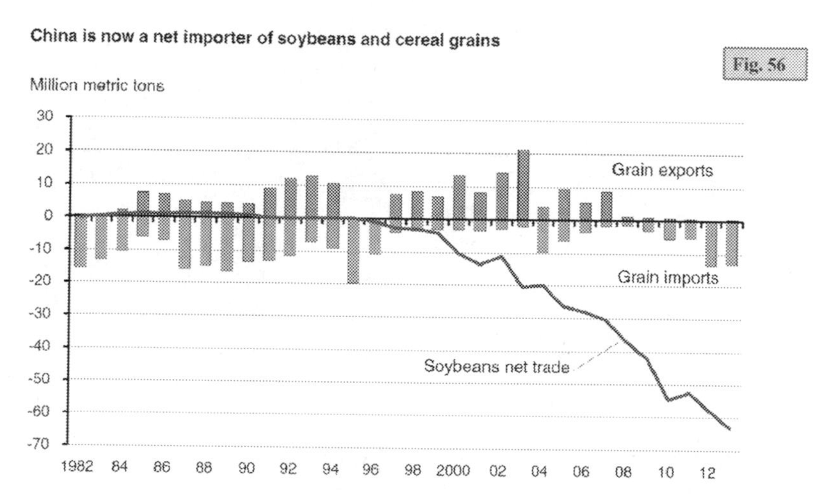

Cereals and soybean import and export by China over 1982 through 2012. Soybean import increased rapidly since the mid 1990's.

With continuing high economic growth in China, the predictions for the years 2023 show continued increase in import of agricultural products, particularly soybean import attaining 112 million metric tons (Table 8). USDA projections based on the rising income of Chinese people tied to changing food habits show that agricultural exports to China will continue to grow in the future. China will continue to be the major buyer of world cotton, since it will remain as the major manufacturer of textiles in the world despite rising wages of the textile workers in China.

Projections of China's imports of selected commodities in 2023

Commodity	USDA	OECD-FAO	China Academy of Agricultural Sciences	China Research Center for Rural Economy
			Million metric tons	
Rice	2.4	2.4	2.3	1.8
Wheat	5.5	2.8	3.1	5.7
Corn	22.0	16.9*	12.0	18.0
Soybeans	112.0	81.5*	74.0	75.0
Meat	2.4	3.6	2.4	NA
Cotton	4.6	3.3	1.6	3.0

*OECD-FAO's corn projection is for all coarse grains; soybean projection is for all oilseeds. NA=Not available.

Notes: OECD/FAO refers to the Organization for Economic Co-operation and Development and the Food and Agriculture Organization of the United Nations.

Sources: USDA, Office of the Chief Economist, 2014; OECD-FAO, 2014; China Academy of Agricultural Sciences, 2014; Xu et al., 2014.

Reproduced with Permission from: USDA – Economic Research Service

Table 8.

Despite rapid growth in China's agricultural imports, the trade with China is complicated by various difficulties and uncertainties, such as often changing and new requirements for exporters, rejection of shipments, antidumping investigations, and lack of consistencies in inspections and quarantine regulations etc.

REFERENCES

1. Alva, A.K., Lanyon, L.E. and Leath, K.T. Excess soil water and Phytophthora root rot stresses of Phytophthora root rot sensitive and resistant alfalfa cultivars. Agron. J. 77:437-442. 1985.
2. Alva, A.K. MN-methoden til Thailand. Ugeskrift for Jord. 34:930-931. 1985.
3. Alva, A.K., Lanyon, L.E. and Leath, K.T. Influence of fungal-soil water interactions on Phytophthora root rot of alfalfa. Biol. Fert. Soils 1:91-96. 1985.
4. Alva, A.K., Edwards, D.G., Asher, C.J. and Blamey, F.P.C. Effects of P/Al molar ratio and calcium concentration on plant response to aluminum toxicity. Soil Sci. Soc. Am. J. 50:133-137. 1986.
5. Alva, A.K., Edwards, D.G., Asher, C.J. and Blamey, F.P.C. Relationships between root length of soybean and calculated activities of aluminum monomers in nutrient solution. Soil Sci. Soc. Am. J. 50:959-962. 1986.
6. Alva, A.K., Asher, C.J. and Edwards, D.G. The role of calcium in alleviating aluminum toxicity. Aust. J. Agric. Res. 37:375-382. 1986.
7. Alva, A.K., Sumner, M.E. and Noble, A.D. Alleviation of aluminum toxicity by phosphogypsum. Commun. Soil Sci. Plant Anal. 19:385-403. 1988.
8. Alva, A.K., Edwards, D.G., Carroll, B.J., Asher, C.J. and Gresshoff, P.M. Nodulation and early growth of soybean mutants with increased nodulation capacity under acid soil infertility factors. Agron. J. 80:836-841. 1988.
9. Alva, A.K. and Sumner, M.E. Alleviation of aluminum toxicity to soybeans by phosphogypsum or calcium sulfate in dilute nutrient solutions. Soil Sci. 147:278-285. 1989.
10. Alva, A.K., Gascho, G.J. and Guang, Y. Gypsum material effects on peanut and soil calcium. Commun. Soil Sci. Plant Anal. 20:1727-1744. 1989.
11. Alva, A.K. and Sumner, M.E. Amelioration of acid soil infertility by phosphogypsum. Plant and Soil. 128:127-134. 1990.
12. Alva, A.K., Sumner, M.E. and Miller, W.P. Reactions of gypsum or phosphogypsum in highly weathered acid subsoils. Soil Sci. Soc. Am. J. 54:993-998. 1990.

13. Alva, A.K. and Singh, M. Effects of soil-cation composition on reactions of four herbicides in a Candler fine sand. Water, Air and Soil Pollut. 52:175-182. 1990.
14. Alva, A.K. and Singh, M. Use of adjuvants to minimize leaching of herbicides in soil. Environ. Manag. 15:263-267. 1991.
15. Alva, A.K. and Sumner, M.E. Characterization of phytotoxic aluminum in soil solutions from phosphogypsum amended soils. Water, Air, Soil Pollut. 57-58:121-130. 1991.
16. Alva, A.K. Differential leaching of nutrients from soluble vs. controlled-release fertilizers. Environ. Manag. 16:769-776. 1992.
17. Zhu, B. and Alva, A.K. Differential adsorption of trace metals by soils as influenced by exchangeable cations and ionic strength. Soil Sci. 155:61-66. 1993.
18. Zhu, B. and Alva, A.K. The chemical forms of Zn and Cu extractable by Mehlich 1, Mehlich 3, and ammonium bicarbonate-DTPA extractions. Soil Sci. 156:251-258. 1993.
19. Alva, A.K. Sorption of calcium and sulfate by sandy soils from flue-gas desulfurization gypsum and phosphogypsum. Commun. Soil Sci. Plant Anal. 24:2701-2713. 1993.
20. Bilski, J.J. and Alva, A.K. Transport of heavy metals and cations in a fly ash amended soil. Bull. Environ. Contam. Toxicol. 55:502-509. 1995.
21. Zhang, M., Alva, A.K., Li, Y.C. and Calvert, D.V. Fractionation of iron, manganese, aluminum, and phosphorus in selected sandy soils under citrus production. Soil Sci. Soc. Am. J. 61:794-801. 1997.
22. Zhang, M., Alva, A.K., Li, Y.C. and Calvert, D.V. Chemical association of Cu, Zn, Mn, and Pb in selected sandy citrus soils. Soil Sci. 162:181-188. 1997.
23. Alva, A.K., Paramasivam, S. and Graham, W.D. Impact of nitrogen management practices on nutritional status and yield of Valencia orange trees and groundwater nitrate. J. Environ. Qual. 27:904-910. 1998.
24. Fares, A. and Alva, A.K. Estimation of citrus evapotranspiration by soil-water mass balance. Soil Sci. 164: 302-310. 1999.
25. Paramasivam, S., Alva, A.K., Prakash, O. and Cui, S.L. Denitrification in Vadose zone and in surficial groundwater of a sandy entisol with citrus production. Plant Soil 208: 307-319. 1999.
26. He, Z.L., Alva, A.K., Calvert, D.V., Li, Y.C. and Banks, D.J. Sorption/desorption and solution concentration of phosphorus at various depths in a fertilized sandy soil. J. Environ. Qual. 28: 1804-1810. 1999.
27. He, Z.L., Alva, A.K., Calvert, D.V. and Banks, D.J. Ammonia volatilization from different fertilizer sources, and effects of temperature, and soil pH. Soil Sci. 164:750-758. 1999.

28. Dou, H., Alva, A.K. and Bondada, B.R. Growth and chloroplast ultrastructure of two citrus rootstock seedlings in response to ammonium and nitrate nutrition. J. Plant Nutri. 22:1731-1744. 1999.
29. Fares, A. and Alva, A.K. Soil water components based on capacitance probes in a sandy soil. Soil Sci. Soc. Am. J. 64:311-318. 2000.
30. Alva, A.K., Paramasivam, S., Graham, W.D. and Wheaton, T.A. Best nitrogen and irrigation management practices for citrus production in sandy soils. Water, Air, and Soil Pollution. 143:139-154, 2003.
31. Mattos, Jr., D., Graetz, D.A. and Alva, A.K. Biomass distribution and 15N Partitioning in citrus trees on a sandy Entisol. Soil Sci. Soc. Am. J. 67:555-563, 2003.
32. Mattos, Jr., D., Quaggio, J.A., Cantarella, H. and Alva, A.K. Nutrient content of biomass components of Hamlin (Citrus sinensis (L.) Osb.) sweet orange trees. Scientia Agricola 60:155-160, 2003.
33. Alva, A.K., Paramasivam, S., Obreza, T.A. and Schumann, A.W. Nitrogen best management practice for citrus trees: I. Fruit yield, quality, and leaf nutritional status. Scientia Horticultureae 109:223-233, 2006.
34. Alva, A.K., Paramasivam, S., Fares, A., Obreza, T.A. and Schumann, A.W. Nitrogen best management practice for citrus trees II. Nitrogen fate, transport, and components of N budget. Scientia Horticultureae. 109:223-233, 2006.
35. Alva, A.K. Sustainable nutrient management in sandy soils - fate and transport of nutrients from animal manure vs. inorganic sources. J. Sustainable Agric. 28:139-155. 2006.
36. Alva, A.K. Setpoints for potato irrigation using real-time continuous monitoring of soil-water content in soil profile. J. Crop Improvement. 21:117-137. 2008.
37. Alva, A.K., Mattos Jr., D. and Quaggio, J.A. Advances in nitrogen fertigation of citrus. J. Crop Improvement. 22: 121-146. 2008.
38. Alva, A.K., Fan, M., Chen, Q., Rosen, C., Ren, H. Improving Nutrient Use Efficiency in Chinese Potato Production – Experiences from the US. Journal of Crop Improvement. 25: 46-85. 2011.
39. Lee, C.H., Liu, G., and Alva, A.K. Potato cultivar evaluation for phosphorus-use efficiency. Journal of Crop Improvement. 27: 617-626, 2013.
40. Alva, A.K., Bilski, J.J., Sajwan, K.S. and van Clief, D. Leaching of metals from soils amended by fly ash and organic by-products. pp. 193-206. K.S. Sajwan et. al. (eds.) Biogeochemistry of Trace Elements in Coal and Coal Combustion Byproducts. Kluwer Academic/Plenum Publishers, New York, NY. 1999. (Book Chapter).

41. Sajwan, K.S., Alva, A.K., and Keefer, R.F. (Eds.) Biogeochemistry of trace elements in Coal and Coal Combustion Byproducts. Kluwer Academic/ Plenum Publishers, New York, NY. p. 359. 1999. (Edited Book)
42. Alva, A.K. Soil pollution. pp. 573-575. In: D.E. Alexander and R.W. Fairbridge (eds.) Encyclopedia of Environmental Science. Kluwer Academic Publishers, Dordrecht, The Netherlands. 1999. (Encyclopedia Article)
43. Alva, A.K. Fertilizers. pp. 249-250. In: D.E. Alexander and R.W. Fairbridge (eds.) Encyclopedia of Environmental Science. Kluwer Academic Publishers, Dordrecht, The Netherlands. 1999. (Encyclopedia Article)
44. Sajwan, K.S., Alva, A.K. and Keefer, R.F. Chemistry of Trace Elements in Fly Ash. Kluwer Academic/Plennum Publishers, New York, NY p. 346. 2003. (Edited Book)
45. Alva, A.K., Paramasivam, S., Fares, A., Delgado, J.A., Mattos Jr., D. and Sajwan, K.S. Nitrogen and irrigation management practices to improve nitrogen uptake efficiency and minimize leaching losses. Goyal, S.S., R. Tischner, and A.S. Basra (eds). In Enhancing the Efficiency of Nitrogen Utilization in Crops. Haworth's Food Products Press, Binghamton, New York. Pp. 369-420, (co-published in Journa l of Crop Improvement, Vol. 15, 2005) 2006. (Book Chapter)
46. Sajwan, K.S., Twardwoska, I., Punshon, T. and Alva, A.K. Coal combustion byproducts and environmental issues. Springer Science and Business Media, Inc., New York, NY. P. 241. 2006. (Book Edited)
47. Alva, A.K., H.P. Collins, R.A. Boydston (2009). Nitrogen Management for Irrigated Potato Production under Conventional and Reduced Tillage. Soil Sci. Soc. Am. J. 73:1496:1503
48. Baumert, K, T Herzog, and J Pershing. 2005. Climate Data: A Sectoral Perspective. World Resources Institute, Pew Center on Global Climate Change. 1 – 35
49. Brentrup, F. 2009. The impact of mineral fertilizers on the carbon footprint of crop production. The Proceedings of the International Plant Nutrition Colloquium XVI, Department of Plant Sciences, UC Davis, UC Davis, 04-09-2009.
50. http://escholarship.org/uc/item/19f2h0p9 Accessed on March 15, 2011.
51. Bronson, KF, AR Mosier. 1991. Effect of encapsulated calcium carbide on dinitrogen, nitrous oxide, methane, and carbon dioxide emissions from flooded rice. Biol Fertil Soils 11:116-120.
52. Burneya, JA, SJ Davis, and DB Lobell. 2010. Greenhouse gas mitigation by agricultural intensification. Proc Natl Acad Sci 107:12052-12057, doi: 10.1073/pnas.0914216107.

53. Chan, KY, L Van Zwieten, I. Meszaros, A Downie, and S Joseph. 2008. Using poultry litter biochars as soil amendments. Australian Journal of Soil Research 46: 437-444.
54. Cole, CV, J Duxbury, J Freney, O Heinemeyer, K Minami, A Mosier, K Paustian, N Rosenberg, N Sampson, D Sauerbeck, and Q Zhao. 1997. Global estimates of potential mitigation of greenhouse gas emissions by agriculture. Nutrient Cycling in Agroecosystems 49: 221–228.
55. Collins, HP, J.L. Smith, S. Fransen, A.K. Alva, C.E. Kruger and D.M. Granatstein. 2010. Carbon Sequestration under Irrigated Switchgrass (Panicum virgatum L.) Production. Soil Science Soc. America 74: 2049-2058.
56. Fastdomain.com. 2010. Environment. http://one-simple-idea.com/Environment1.htm
57. EPA. 2011. Inventory of U.S. Greenhouse Gas Emissions and Sinks: 1990-2008. EPA 430-R-10-006 http://www.epa.gov/climatechange/emissions/usgginventory.html Accessed on March 15, 2011.
58. EPA, 2006. U.S. Greenhouse Gas Inventory Report, April 2006 http://yosemite.epa.gov/opa/admpress.nsf/bb1285e857b49ac4852572a00065683f/7510b703526bc37b85257153006e5add!OpenDocument Accessed on March 15, 2011.
59. EPA. 2011. 2011 Draft U.S. Greenhouse Gas Inventory Report. http://epa.gov/climatechange/emissions/usinventoryreport.html
60. Gale, F., Hansen, J., Jewison, M. 2015. China's Growing Demand for Agricultural Imports. USDA, Economic Research Service, Economic
61. Information Bulletin Number 136, February 2015.
62. Haile-Mariam, S, HP Collins, and SS Higgins. 2008. Greenhouse Gas Fluxes from an Irrigated Sweet Corn (Zea mays L.)–Potato (Solanum tuberosum L.) Rotation. Journal of Environmental Quality 2008 37: 3: 759-771.
63. Hansen, J, M Sato, R Ruedy, K Lo, DW Lea, and M Medina-Elizade. 2006. Global temperature change. Proc Natl Acad Sci 103: 14288-14293, doi: 10.1073/pnas.0606291103
64. Coeli M. Hoove. 2003. Soil Carbon Sequestration and Forest Management: Challenges and Opportunities In: Kimble, JM, LS Heath, RA Birdsey, and R Lal. (eds.). The Potential of U.S. forest soils to Sequester Carbon and Mitigate the Greenhouse Effect. CRC Press LLC, Boca Raton, Florida. Pp. 231-238
65. International Energy Agency (IEA), 2011. Technology Roadmap. Biofuels for Transport. © OECD/IEA, 2011, International Energy Agency, 9 rue de la Fédération 75739 Paris Cedex 15, France, www.iea.org, pp

56. http://www.iea.org/publications/freepublications/publication/biofuels_roadmap_web.pdf
66. Intergovernmental Panel on Climate Change. 2001. Climate Change 2001: The Scientific Basis. Intergovernmental Panel on Climate Change, JT Houghton, Y Ding, DJ Griggs, M Noguer, PJ van der Linden, X Dai, CA Johnson, and K Maskell (eds). Cambridge University Press. Cambridge, United Kingdom. http://www.grida.no/publications/other/ipcc_tar/?src=/climate/ipcc_tar/wg1/001.htm Accessed on March 15, 2011.
67. IPCC. 2007. Climate Change 2007: The Physical Science Basis. Contribution of Working Group I to the Fourth Assessment Report of the Intergovernmental Panel on Climate Change. S Solomon, D Qin, M Manning, Z Chen, M Marquis, KB Averyt, M Tignor, and HL Miller (eds). Cambridge University Press. Cambridge, United Kingdom, 996 pp.
68. Keerthisinghe, DG, JR Freney, and AR Mosier. 1993. Effect of wax-coated calcium carbide and nitrapyrin on nitrogen loss and methane emission from dry-seeded flooded rice. Biol Fertil Soils 16:71-75.
69. Li, C. 2007. Quantifying greenhouse gas emissions from soils: Scientific basis and modeling approach. Soil Science and Plant Nutrition 53: 344–352.
70. McLaughlin, S.B., and Walsh, M.E. 1998. Evaluating environmental consequences of producing herbaceous crops for bioenergy. Biomass and Bioenergy. 14:317-324
71. Marrin, D.L., Reducing Water and Energy Footprints via Dietary Changes among Consumers. International Journal of Nutrition and Food
72. Sciences. Vol. 3, No. 5, 2014, pp. 361-369. doi: 10.11648/j.ijnfs.20140305.11
73. Milich, L. 1999. The role of methane in global warming: where might mitigation strategies be focused? Global Environmental Change 9: 179-201.
74. Mosier, AR, JM Duxbury, JR Freney, O Heinemeyer, and K Minami. 1996. Nitrous oxide emissions from agricultural fields: Assessment, measurement and mitigation. Plant and Soil 181: 95-108.
75. NOAA/ESRL. 2009. Trends in Atmospheric Carbon Dioxide. Available online at http://www.esrl.noaa.gov/gmd/ccgg/trends/. 11 January 2010.
76. Paustian, K, J Six, ET Elliott, and HW Hunt. 2000. Management options for reducing CO2 emissions from agricultural soils. Biogeochemistry 48: 147–163.
77. Sauerbeck, DR. 2001. CO2 emissions and C sequestration by agriculture – perspectives and limitations. Nutrient Cycling in

Agroecosystems 60: 253–266. Schneider, UA and BA McCarl. 2003. Economic Potential of Biomass Based Fuels for Greenhouse Gas Emission Mitigation. Environmental and Resource Economics 24: 291–312.

78. Schneider, UA and BA McCarl. 2003. Economic potential of biomass based fuels for greenhouse gas emission mitigation. Environmental and Resource Economics 24: 291312.
79. Scott, HD and FG Renaud. 2007. Aeration and Drainage. In: Irrigation of agricultural crops (2nd ed). RJ Lascano and RE Sojka (eds). 195-235. ASA-CSA-SSSA, Inc. Madison, WI.
80. Tampier, M., Smith, D., Bibeau, E. and Beauchemin, P. 2004. Identifying environmentally preferable uses for biomass resources, Stage 2 report: Life-cycle GHG emission reduction benefits of selected feedstock-to-product threads. Prepared for Natural Resources Canada and the National Research Council of Canada by Envirochem Services Inc, North Vancouver, British Columbia
81. Tilman, D, J Hill, and C Lehman. 2010. Carbon-Negative Biofuels from Low-Input High-Diversity Grassland Biomass. Science. 314: 1598-1600.
82. Towprayoon, S. 2004. Greenhouse Gas Mitigation Options from Rice Field. Presented at In-session workshop on Climate Change Mitigation 19 Bonn 2004, Maritim Hotel, Bonn. http://docs.google.com/viewer?a=v&q=cache:yzZi4_DtKaQJ:unfccc.int/files/meetings/workshops/other_meetings/application/vnd.ms-powerpoint/towprayoon.ppt+Towprayoon,+2004,+Greenhouse+Gas+Mitigation+Options+from+Rice+Field.&hl=en&gl=us&pid=bl&srcid=ADGEESj5ZLbQMXCVrD45n1oUYNV5d7KgJCNonMmErtvr9ORCNIF5I5v0yrgqJbLEXWDoAwOHYewHZG5ywFm5KLboml15wyWR2anoSMgg1BOMP3UHreJF9Fx_Z0hBsoCgRZZF-itsUZtt&sig=AHIEtbRnWr-2CsOqnsVES8o8QTX3epTyxw&pli=1 . Accessed on March 15, 2011.
83. Wang, Q, Y. Li, AK Alva. 2010. Growing Cover Crops Improve Biomass Accumulation and Carbon Sequestration: A Phytotron Study. Journal of Env Protect 1:73-84.
84. USDA – Economic Research Service. 2010. Food dollar, Report Series.
85. Weinheimer, J N Rajan, P Johnson, and S Maas. 2010. Carbon Footprint: A New Farm Management Consideration in the Southern High Plains. Agricultural & Applied Economics Association 2010. Denver, Colorado, July 25-27, 2010 http://ageconsearch.umn.edu/bitstream/61760/2/JustinWeinheimer2010AAEA.pdf Accessed on March 15, 2011.

86. West, TO and G Marland. 2002. A synthesis of carbon sequestration, carbon emissions, and net carbon flux in agriculture: comparing tillage practices in the United States. Agric Ecosyst Enviorn 91: 217-232.
87. Woolf, D, JE Amonette, FA Street-Perrott, J Lehmann, and S Joseph. 2010. Sustainable biochar to mitigate global climate change. Available online at http://www.nature.com/ncomms/journal/v1/n5/pdf/ncomms1053.pdf Accessed on March 15, 2011.
88. One Simple Idea. 2011. Environment. http://one-simple-idea.com/Environment1.htm Accessed on March 15, 2011
89. National Oceanic & Atmosphere Administration Earth System Research Laboratory. 2011. Trends in Atmospheric Carbon Dioxide. http://www.esrl.noaa.gov/gmd/ccgg/trends/ Accessed on March 15, 2011
90. Butler, R.A. 2006. High oil prices fuel bioenergy push. MONGABAY. http://news.mongabay.com/2006/05/high-oil-prices-fuel-bioenergy-push/
91. Monteiro, N., and I. Altman. 2012. The impact of ethanol production on food prices: The role of interplay between the U.S. and Brazil. Energy Policy 41:193-199.

Printed in the United States
By Bookmasters